To Lauri,
with eternal thanks
for joining this team.

Leadership in SPACE

Leadership in SPACE

Selected Speeches of NASA Administrator Michael Griffin, May 2005–October 2008

National Aeronautics and Space Administration

Headquarters
300 E St SW
Washington, DC 20546 2008

www.nasa.gov

NASA SP-2008-564

Library of Congress Cataloging-in-Publication Data

Griffin, Michael D. (Michael Douglas), 1949-
Leadership in space : selected speeches of NASA administrator Michael Griffin, May 2005-October 2008 / Michael D. Griffin.
p. cm.
"NASA SP-2008-564."
1. Astronautics—United States. 2. Astronautics and state—United States. 3. Griffin, Michael D. (Michael Douglas), 1949- 4. United States. National Aeronautics and Space Administration—Management. 5. Competition, International. 6. Outer Space—Exploration—Government policy—United States. I. United States. National Aeronautics and Space Administration. II. Title.
TL789.85.G68A5 2008
629.4092—dc22

2008039636

For sale by the Superintendent of Documents, U.S. Government Printing Office
Internet: bookstore.gpo.gov Phone: toll free (866) 512-1800; DC area (202) 512-1800
Fax: (202) 512-2104 Mail: Stop IDCC, Washington, DC 20402-0001

ISBN 978-0-16-081565-2

Table of Contents

Part 3.
Getting There from Here

Preface
Leadership in Space

Section 203 (a) (3) of the National Aeronautics and Space Act of 1958 requires of NASA that, "The Administration, in order to carry out the purpose of this Act, shall provide for the widest practicable and appropriate dissemination of information concerning its activities and the results thereof." This clause provides the legal basis for all of NASA's public outreach activities, among which are included the many speeches given by the administrator and other senior officials in the course of representing the agency. While public speaking does not come naturally to me, it is an important part of my duties; and from the outset of my tenure, I determined to abide by several tenets I thought to be important when representing NASA in public forums.

First, I thought that each speech should be unique. I wanted each new speech to have something "fresh" for listeners, some new food for thought. While there would of course be common themes I wanted to emphasize, I would try to do so in different ways at each venue in which I spoke. Second, each speech would be "mine" by the time I gave it—my own thoughts in my own "voice." I was willing, indeed eager, to accept thematic ideas, drafts or editing suggestions from any source. But I didn't intend to use a "speechwriter" in the conventional sense. By the time a speech was ready to be delivered, I would have reworked it so thoroughly that I would be at least a co-author in the fullest sense. I would never simply read a speech prepared by others, even if the alternative was to deliver an extemporaneous address on the spot, as sometimes happened. And finally, while each speech was to be unique, I wanted to leave behind my thoughts on certain themes of particular interest to me and whose treatment would be the topic of more than one address.

The most important theme for me has been that of identifying and elucidating an intellectual rationale for human space exploration. I believe that it is time for the proponents of human spaceflight to cease the attempt to justify it on the basis of its contributions to science. While human spaceflight does offer options and venues for scientific discovery that are not easily provided by robotic spacecraft, this is merely a collateral benefit of human presence in space.

Human space exploration is not fundamentally about enabling science, important as that is. It is about expanding the range of human action, experience and influence. It is about developing options for succeeding generations to exploit the resources of the solar system for the benefit of mankind. I believe that this is an endeavor whose intrinsic merit stands on par with that of scientific discovery in its value to the human race. Yet when I joined NASA in the year following President George W. Bush's announcement of the Vision for Space Exploration, comments such as "exploration without science is tourism" were being bandied about in the media. Such comments convey the disdainful judgment that human exploration exists, if it is allowed to exist at all, on the sufferance of the scientific community and is to be ranked in terms of its value to that community. Even the International Space Station, the most incredibly stunning engineering achievement in human history, was justified in terms of the arcane science it would enable instead of its truly fundamental role as a toehold on a new frontier, a place to begin to learn how to live and work in space.

I sought to offer through my public addresses a more complete rationale for human presence in space. In various speeches I tried to illustrate the connections between a vigorous program of human space exploration and important societal interests including national security; the value of leadership and partnership in great enterprises; the effect on our industrial base and economic competitiveness of learning how to do very hard things; the spread of values and culture among the worlds of the future; the long-term survival of the human species; and, yes, enhanced possibilities for scientific discovery. Part 1 of this collection offers a collection of speeches addressing various aspects of that broad theme.

Because the exploration and development of the space frontier is a strategic issue for the United States, NASA touches important segments of society on many levels. Learning how to live and work in space or sending our machines there on our behalf has profoundly influenced and in some cases redefined the practice of commerce, science, engineering and management in our nation and the world. Part 2 includes speeches on various aspects of such influence. In particular, I am persuaded that we owe the modern practice of system engineering and its allied discipline, systems management, to the driving force of the early civil and military space programs. I believe these disciplines are crucial to our success in a competitive global society yet are widely misunderstood; and so I have included in this section some material outlining my views on what system engineering is and is not.

We are at a cusp of opportunity for the U.S. civil space program. Apollo, and with it the outward focus of our efforts in space, was ended almost before the policy makers of that era had time to consider or debate the implications of what was being done and not done. Spotlighted by the searching examination of the *Columbia* Accident Investigation Board was the glaring assessment that the U.S. space program had moved forward for more than three decades without a guiding vision. For a brief moment, and for the first time in decades, this was broadly seen to be the fundamental flaw in national space policy that some of us have always seen (and lamented) but were powerless to alter. But out of that tragedy came a new guiding vision, one that I believe is perfectly aligned with this nation's proper interests in space.

Now, at this cusp, we who want this new vision carried forward must recognize that it is not yet cemented firmly in place. Outward, destination-focused exploration is not yet the paradigm for "what NASA does." For decades, the public image of NASA has been that of flying the space shuttle to orbit, carrying out activities little understood and less regarded by most of those who must support them. As it retreated into history, Apollo assumed mythological proportions, becoming simultaneously NASA's apotheosis and a measure of what had been lost. The actual doing of Apollo seemed, in early 2005, to be impossibly beyond us. Even now, very few in our space industry yet believe that we can really, truly, actually, not only re-create but go beyond the achievements of the Apollo years.

But I know that we can. And Part 3 is devoted to my assessment, in rather specific terms, of why I believe that is so, how we can do it and how what we are doing fits into the context of what we have done before. The civil space strategy that has been laid out is practically achievable and affordable, if barely so, with the budgets we can expect to receive. I hope that you will conclude after reading the material in this section that this is indeed so.

So, finally, in the work presented here I have tried to provide a thoughtful rationale for what we are doing in space, why we are doing it and how we intend to bring it about. If we can agree, as a community of those who believe that our future lies in mastering the space frontier, then I hope you will also agree with me that now it is a future we can seize.

Part 1.
Exploration and Our Future

Leadership in Space

Michael D. Griffin
Administrator
National Aeronautics and Space Administration

California Space Authority
December 2, 2005

I'm here today to talk about national and world leadership in space—what it means to me, and what I think it takes to achieve and maintain it.

Most will agree that it is important for the United States to be a leader among the nations of the world, and that such leadership has many dimensions. Economic, cultural, diplomatic, moral and educational leadership are certainly major components of world leadership. Moreover, we clearly still live in a time when any wealthy and prominent nation must have the ability to defend itself and its allies. But true leadership also involves defining, and then pursuing, the frontiers that expand mankind's reach. It means occupying the cutting edge of science and technology. It means establishing world technical standards—as we have done in the computing and aviation industries—not through coercion but because we have developed a capability that others wish to use. It also means having the ability and determination to take the lead in building coalitions and partnerships to do those things that fulfill the dreams of mankind. And those dreams have always included the desire to see what lies beyond the known world.

To journey beyond the known world today, we must leave Earth entirely. That is the long-held dream that has actively engaged our country and others for nearly 50 years, since our first primitive steps in the exploration of space became possible. And I firmly believe that in the 21st century taking shape as we speak, a vital part of world leadership will be leadership in the exploration and development of the space frontier.

For many years, our country has been rightly recognized as the world leader in the exploration and use of space, and in developing and deploying the technologies that make space leadership possible. Our determination to be first on the moon and preeminent in other space activities resulted in some of the iconic moments of the 20th century, and helped to solidify American leadership in the generation after World War II.

But, as they say, that was then and this is now. We cannot rest on, nor be satisfied with, past accomplishments. The true space age, in which humans will explore the worlds beyond our own, is just getting underway. Leadership in establishing a human presence in the solar system will, in my judgment, be a key factor in defining world leadership on Earth for generations to come.

Throughout history, the great civilizations have always extended the frontiers of their times. Indeed, this is almost a tautology; we define as "great" only those civilizations which did explore and expand their frontiers, thereby ultimately influencing world culture. And when, inevitably, some societies retreated from the frontiers they had pioneered, their greatness subsided as well.

Today, other nations besides our own aspire to leadership on the space frontier. These nations are making progress, and they will undoubtedly utilize their advancements in space to influence world affairs. Their activities will earn them the respect, which is both sincere and automatic, that is accorded to nations and societies engaged in pioneering activities. These things are not in doubt, and so the question before us is this: when other nations reach the moon, or Mars, or the worlds beyond, will they be standing with the United States, or will we be watching their exploits on television? The president has given us his answer. America will lead. In 2004, the president said, "We have undertaken space travel because the desire to explore and understand is part of our character. And that quest has brought tangible benefits that improve our lives in countless ways." He also said our Vision for Space Exploration is a "journey, not a race." These words are unambiguous. They chart a course for action that is unmistakable. It is imperative that this commitment transcend any given administration and any given Congress.

Today, as other countries renew their commitment to space, America has the opportunity, and I would argue the obligation to maintain our leadership role in space exploration. As we watch other countries commit to developing new exploration systems and technologies to expand into space, we too must remain committed to new advancements, lest we fall behind. In that regard, it may be significant to note that of today's major spacefaring powers, only Russia and China have spacecraft—*Soyuz* and *Shenhzou*—that are capable of returning crews from a trip to the moon.

Through the Vision for Space Exploration however, this country has a renewed commitment to maintain our leadership and restore the capabilities we set aside many years ago. The vote by three successive Congresses to support the Vision for Space Exploration outlined by President Bush in 2004 offers wonderful evidence of national determination to regain lost ground in space. But beyond those very important congressional votes, there are some very serious challenges that we must face as a nation. We must think carefully about what the world of tomorrow will look like if the United States is not the preeminent spacefaring nation. And if we don't like that picture, if we truly want the United States to be the world leader in space now and in the future, there are a number of critical things we simply must decide to do. The Vision gives us the opportunity to take on the leading role in the exploration of space, not just for this century, but for centuries to come. But we have to seize that opportunity and make it a reality.

The first essential step is that American leadership in the exploration and development of the space frontier must be an explicit national goal. There must be continued and sustained bipartisan cooperation and agreement on the importance and necessity of American leadership in space, just as we are determined to be leaders in other areas such as defense, education and scientific research. There need not, indeed there must not, be partisan debates over whether to have a vibrant space program or not. And we must get beyond revisiting this determination each year or after an accident, or after a technical problem.

In addition to needing national agreement on the importance of American leadership in space, we need to make this a commitment from generation to generation. Space exploration by its very nature requires the planning and implementation of missions and projects over decades, not years. Decades of commitment were required to build up our network of transcontinental railroads and highways, as well as our systems for maritime and aeronautical commerce. It will not be quicker or easier to build our highways to space, and the commitment to do it must be clear and sustaining.

To ensure the success of the space program across a wide spectrum of political thought and down the generations, it is essential to have simple but compelling goals. The space community has an obligation to communicate to the country our plans to ensure America's leadership in space exploration. The President's Vision for Space Exploration has established goals that people can understand and support—moving our space exploration activities beyond low Earth orbit, and returning to the moon as a stepping-stone to Mars and other destinations beyond, such as the near-Earth asteroids.

Broad support for these goals is certainly there. A recent Gallup poll indicated that, with funding levels at or below one percent of the federal budget, three-quarters of Americans are supportive of our plans to return to the moon and voyage to Mars. This is amazingly strong support for any government initiative, and I believe it provides a firm foundation upon which to build in the years ahead. The first step might be to explain that we're actually spending less than 0.7 percent of the federal budget!

Still another key requirement for long-term leadership in space is the ability to build and maintain a strong international coalition of spacefaring nations. A critical component of this ability will always be our credibility in making agreements and honoring them. In any partnership, the most critical commitments fall upon the senior partner. Since that, of course, is the role we wish to play, we must be thoughtful, deliberate and sure about any commitments we make. But once made, we need to keep them. I think we can all agree that one of the best results of the International Space Station program is the cooperation it has fostered among the participating nations. A prime goal of the President's Vision for Space Exploration is to continue and expand this cooperation as we plan for human lunar return.

These are some of the key things we need to do if we Americans are indeed serious about being a leader on the space frontier. As we lift our eyes to the future, I see a space program that will bring hope, opportunity and tangible benefits as we renew our commitment to lead in these endeavors. While we cannot predict today at what pace others will venture beyond Earth orbit and establish the first outposts on distant worlds I earnestly believe those nations that are the most adept at reading the lessons of history will be taking the lead.

I have mused often upon these lessons, looking for the patterns that can provide guidance for our own time. Indeed, if we were alive 500 years ago, or thereabouts, and a candlelight conference were held in Lisbon by the Portuguese Oceans Authority, no doubt we would be listening to such giants of exploration as Vasco da Gama and Pedro Alvares Cabral (the explorer who claimed Brazil for Portugal) explain how their activities would bring about Portugal's rise to global influence.

Perhaps all of us would be speaking Portuguese today had not first Spain, and then later England, made a greater commitment to the discovery, exploration and settlement of new territories.

As an example of how the choices that nations make matter, not only for themselves, but also for the future of humanity, let us consider the case of John Cabot. Cabot, whose true name was Giovanni Caboto, was an Italian who sailed for the English government and private merchants after Spain and Portugal expressed no interest in his ideas on finding a westward passage to Asia. While exploring the coastal regions of North America in Newfoundland, he established the basis for England's claim to North America and was the first to bring our language to the shores where we now live.

There are more recent examples of similar pivotal crossroads in our history. While American ingenuity, in the form of those quintessentially American inventors, Wilbur and Orville Wright, did lead the way into the era of powered flight, we tend to forget that we squandered our initial leadership in aviation. And so, 90 years ago, the National Advisory Committee for Aeronautics, NASA's major predecessor, was founded precisely because our nation's leaders feared the European nations already had a significant advantage in the development of strategically important aviation systems and technologies, just one decade into the age of flight. This was in fact true, and as a consequence, the air war of World War I was fought with European airplanes.

But because we made a strong commitment at that time to this emerging field, the influence of American air power and aviation technology can, today, be seen in everything from the fact that we live in a world not dominated by fascism or communism, to the fact that when you fly anywhere in the world, say from Bangalore to Bangkok, the International Civil Aviation Organization dictates that pilots and air traffic controllers speak English. This is a lesson that cannot be learned too thoroughly: if we become complacent, other nations can and will surpass our achievements.

As we look forward to the events that will define the 21st century, as viewed by the historians of yet future centuries, there is no doubt that the expansion of human civilization into space will be among the great achievements of this era. We have the opportunity, and I would say the obligation, to lead this enterprise, to explore worlds beyond our own and to help shape the destiny of this world for centuries to come.

I am convinced that leadership in the world of the 21st century and beyond will go to the nation that seeks to fulfill the dreams of mankind. We know what motivates those dreams. Exploring new territory when it becomes possible to do so has defined human striving ever since our remote ancestors migrated out of the east African plains. The human imperative to explore new territories, and to exploit the resources of these territories, will surely be satisfied by others if not by us. What the United States gains from a robust, focused program of human and robotic space exploration is the opportunity to define the course along which this human imperative will carry us.

The Vision for Space Exploration affords the United States nothing less than the opportunity to take the lead, not only in this century but in the centuries to follow, in advancing those interests of our nation that are very much in harmony with the interests of people throughout the world. Space will be explored and exploited by humans. The question is: which humans and from where, and what language will they speak? It is my goal that Americans will be always among them. If this is the future we wish to see, we have a lot of work to do to sustain the Vision which takes us there. To me, the choice could not be more compelling.

Space Exploration: Real Reasons and Acceptable Reasons

Michael D. Griffin
Administrator
National Aeronautics and Space Administration

Quasar Award Dinner
Bay Area Houston Economic Partnership
January 19, 2007

We have a very interesting conundrum at NASA, and we have been spending a lot of time lately thinking about it. In national polling, NASA as an American institution enjoys a hugely positive approval rating, broadly in the range of 65 to 75 percent, an amazing result for a government agency. But when you ask people why, they are not really sure, or at least cannot express it clearly. When you ask people what we do beyond the broad category of "space," again they aren't quite sure. And if you ask them what we're planning to do, they're even less sure. But they know that they love NASA. So NASA has what in the marketing discipline would be called "very strong brand loyalty," even though people are not familiar in detail with what we do or why they like it.

I have been trying to understand why this is so, because it is important to our agency's future. If we don't have public support that is both strong and specific, the things we want to do and believe to be important will not survive. There are many competing priorities for public funding, and there always will be. So it really is important for us to communicate to the public how we're spending the 15 cents per day that the average American contributes to NASA, because there are other places where that money can go.

I've reached the point where I am completely convinced that if NASA were to disappear tomorrow, if the American space program were to disappear tomorrow, if we never put up another Hubble, if we never put another human being in space, people would be profoundly distraught. Americans would feel less than themselves. They would feel that our best days are behind us. They would feel that we have lost something, something that matters. And yet they would not know why.

This is an interesting conclusion, and so I've thought about it a good bit; and I've come believe that the reason is that we in the space business don't talk about it in the right way.

If you ask why we're going back to the moon and, later, beyond, you can get a variety of answers. The president quite correctly said that we do it for purposes of scientific discovery, economic benefit and national security. I've given speeches on each of those topics, and I think these reasons can be clearly shown to be true. And Presidential Science Advisor Jack Marburger has said that questions about space exploration come down to whether or not we want to bring the solar system within mankind's sphere of economic influence. I think that is extraordinarily well put.

These reasons have in common the fact that they can be discussed within the circles of public policy making. They can be debated on their merits, on logical principles. They can be justified. They are what I am going to call tonight "Acceptable Reasons." You can attach whatever importance you want to any of those factors, and some citizens will weight some factors more and some will weight them less, but most of us would agree that they are, indeed, relevant factors.

But who talks like that? Who talks about doing something for purposes of scientific or economic gain or national security other than in policy circles? If anybody asked Lindberg why he crossed the Atlantic—and many did—he never indicated that he personally flew the Atlantic to win the Orteig prize. His backers might have done it in part for that, but Lindberg did it for other reasons.

If you ask Burt Rutan why he designed and built *Voyager* and why Dick Rutan and Jeanna Yeager flew it around the world, it wasn't for any money involved; it was because it was one of the last unconquered feats in aviation. If you ask Burt and his backer Paul Allen why they developed a vehicle to win the X-Prize, it wasn't for the money. They spent twice as much as they made.

I think we all know why people do some of these things. They are well-captured in many famous phrases. When Sir George Mallory was asked why he wanted to climb Mount Everest, he said, "Because it is there." He didn't say that it was for economic gain.

We know these reasons, and tonight I will call them "Real Reasons." Real Reasons are intuitive and compelling to all of us, but they're not immediately logical. They're exactly the opposite of "Acceptable Reasons," which are eminently logical but neither intuitive nor emotionally compelling. The Real Reasons we do things like exploring space involve competitiveness, curiosity and monument building. So let's talk about them.

First, most of us want to be, both as individuals and as societies, the first, the best and the most, in at least some activity. We want to stand out. This kind of behavior is rooted in our genes. We are today the survivors of people who wanted to outperform others. Without question, that can be carried to an unhealthy degree; we've all seen more wars than we like. But because this trait can be taken too far doesn't mean that we can do without it completely. Competitiveness is rooted in the genes of successful people.

As to curiosity, who among us does not know the wonder, mystery, awe and magic of seeing something—even on television—never seen before, an experience brought back to us by a robotic space mission? And how much grander when one of our own, a representative of other human beings, is there to see it for herself? Who doesn't know that feeling? The urge to know what's over the next hill is one of the most common feelings we share, whatever our backgrounds.

We like to do what I'll call "monument building." We want to leave something behind for the next generation, or the generations after that, to show them that we were here, to show them what we did with our time here. This is the impulse behind cathedrals and pyramids and many, many other things. We could have done a lot of different things to honor George Washington. But what was done was that, in the early 1800s, people started to work on a 550-foot high obelisk to honor him.

But it is not only George Washington whom the monument honors; it says fully as much about the people who built it. And that's okay. It is my observation that when we do things for Real Reasons as opposed to Acceptable Reasons, we produce our highest achievements. The people who do things for Real Reasons, and who know it, are also the ones who are the most successful by the standards embodied in Acceptable Reasons.

All of you in the space business know this, whether you realize it or not, because none of us is in this business for the money to be made. But I believe we see it most obviously in our society and in sports. In my own sport, golf, certain people have, over the decades, risen to the very top of the game and stayed there. People like Bobby Jones, Ben Hogan, Jack Nicklaus or, today, Tiger Woods. In other sports, people like Wayne Gretzky or Michael Jordan come to mind.

What do these people have in common and what is the lesson for the rest of us? The lesson is that they became legends because they wanted to be the very best at what they do. They wanted to leave something behind them, lasting records in their sport. And they wanted to do it because the challenge was there. Who thinks that any of them played, or kept playing, for the money?

I think that tells us something. When you do things for Real Reasons instead of Acceptable Reasons, you have a chance to obtain Real Success. And so we have a conundrum. The cultural ethos in America today requires us to have Acceptable Reasons for what we do. We must have reasons that pass analytical muster, that offer a favorable cost/benefit ratio and that can be logically defended. We tend to dismiss out of hand reasons that are emotional, or are value-driven in ways that we can't capture on a spreadsheet. But, Acceptable Reasons alone don't take us where we really want to go.

In my view, the space business, more than most other endeavors, suffers from the fact that the most important, the best and the most basic reasons for doing it are Real Reasons and not Acceptable Reasons. The Acceptable Reasons—economic benefit, scientific discovery and national security—are, in fact, completely correct. But they comprise a derived rationale and are not the truly compelling reasons. And again, who talks like that, about anything that really matters to them?

Why in today's culture do we focus so much on requiring Acceptable Reasons? Only a couple of generations ago, it was not so much this way.

One observation I would make is that, in the shaping of policy, the kinds of things I've cited as Real Reasons are "right-brain" things; they're intuitive, subjective and difficult to quantify. And they are running around loose in a left-brain world! All of us here tonight got where we are by being analytical and objective and very left-brain oriented. Spaceflight cannot be successfully accomplished without these traits. And so I think we tend not to pay appropriate respect to the deeper parts of human nature that are intuitive and qualitative. This one-sided focus isn't always to our benefit. In a very important sense, we're not the right people to make the arguments as to why we should be encouraged to do what we do!

Some of you must have read Norman Mailer's book from 1970, entitled "Of a Fire on the Moon." Mailer was a unique and controversial novelist. I think of him, in the sense that I was just talking about, as quite possibly the ultimate right-brain kind of guy. And he wrote about Apollo in a very, very interesting book, but from a perspective I've not seen another writer choose. He didn't write about the engineering of it, the operational aspects, the astronauts who flew the missions or anything like that. He wrote about what people were feeling, the power and majesty of the event and the nature of the people who would engage in such a thing. It's a compelling story, but it is not like any other book about the space program that you will read. That's the kind of person, the kind of work, that we need to exemplify the Real Reasons for what we do.

Real Reasons are not amenable to cost/benefit analysis. I'm reminded of the famous quote "A cynic is a man who knows the price of everything and the value of nothing," by the character Lord Darlington in Oscar Wilde's play "Lady Windermere's Fan." It's one of my favorites. Well, in today's America, it's smart; it's popular; it's clever to be a cynic. And a certain amount of it is appropriate; a healthy skepticism of bold claims is necessary. But too much skepticism causes us to deny a part of what we are.

Real Reasons are old fashioned. How many of us grew up reading "Tom Swift," "Jack Armstrong" or "All American Boy"? Or other similar books stories? Not great literature, for sure, but they exemplified many of the values I think we like to see in people: inventiveness, competitiveness, boldness and a sense of good feeling about what it was to be an American, in very simplistic ways, ways which hit close to home.

To read those books was to understand, even as a child, that achievement is to be valued and is not something to be set aside. So, how do we talk about our achievers today? Other than in the field of sports, we talk about today's achievers as "geeks" and "workaholics." People are advised to lead "balanced lives." I don't know about you, but I haven't led a balanced life. But people who want to accomplish something are not balanced. And they are geeks and workaholics. I think we owe our country to people who were like that. I don't know that one could say that folks like George Washington and Thomas Jefferson led balanced lives. Any rational cost/benefit analysis would tell you to stay out of a quarrel with the mother country and let other people deal with it! Who today would talk about pledging "their lives, their fortunes and their sacred honor" to a cause? Today we are uncomfortable with such value discussions, and I think it's a shame.

Now, I talked earlier about building monuments, and I mentioned the cathedrals and the pyramids. Cathedral builders knew what I am talking about tonight. They knew the awe and the mystery of their God. They built monuments to him, and also to themselves, just as the Washington Monument speaks to the people who built it as well as to the person for whom it was built. But they wanted to build the best cathedrals, and if you study cathedral building from a civil engineering perspective, you can see the evolution of that discipline; and you will be impressed. You should be.

Within the space business, Kennedy is probably best remembered for his "Man, Moon, Decade" speech (which, by the way, is also a classic of program management). And it's a great speech. But the Kennedy quote about space that I love more than anything in the world because it evokes exactly the things I'm talking about here tonight, was the one he gave from this lecture at Rice University in September of 1962 when he said, "We choose to go to the moon, and to do the other things, not because they are easy, but because they are hard." I'll say it again: "not because they are easy, but because they are hard."

The cathedral builders knew that reason. They were doing something that required a far greater percentage of their gross domestic product than we will ever put into the space business, and they knew it was hard. We know it too. We look back across 600 or 800 years of time and we are still awed by what they did. What is it that Americans make sure to see when they go to Europe? Who goes to Europe and does not, at some point, see the cathedrals? We are still awed across the centuries by what they accomplished.

To me, the irony is that when we do hard things for the right reasons—for the Real Reasons—we end up actually satisfying all the goals of the Acceptable Reasons. And we can see that, too, in the cathedrals, if we look for it.

What did the cathedral builders get? They didn't just build cathedrals and then stop there. They began to develop civil engineering, the core discipline for any society if it wishes to have anything more than thatched huts. They learned how to build high walls and have them stand up straight. They learned how to put a roof across a long span. They learned which materials would work and which ones would not. And by finding the limits on how high walls could be, how broad roof spans could be and what materials wouldn't work, they created the incentive to solve those problems; so that they could build things beyond cathedrals; so that they could, fundamentally, build Western civilization.

They gained societal advantages that were probably even more important than learning how to build walls and roofs. They learned to embrace deferred gratification, not just on an individual level where it is a crucial element of maturity, but on a societal level where it is equally vital. The people who started the cathedrals didn't live to finish them; such projects required decades. The society as a whole had to be dedicated to the completion of those projects. To be able to do that for cathedrals was to be able to do it in other areas as well. We owe Western civilization as we know it today to that kind of thinking: the ability to have a constant purpose across years and decades.

The medieval builders formed guilds, establishing professional trades beyond that of agriculture. Now, agriculture is at the root of human technology. Nothing good happens to human beings without getting beyond the hunter-gatherer stage, and agriculture is that first step. But the second step is to be able to build physical works that didn't previously exist. The organization and systemization of that in Western society today began in medieval Europe, with the cathedral builders. They learned how to organize large projects, a key to modern society. And, probably most important of all, the cathedrals had to be, for decades at a time, the focus of civic accomplishment and energy. A society, a nation, a civilization, needs such foci.

It is my contention that the products of our space program are today's cathedrals. The space program addresses the Real Reasons why humans do things. It satisfies the desire to compete, but in a safe and productive manner rather than in a harmful manner. It speaks abundantly to our sense of human curiosity, of wonder and awe at the unknown. Who doesn't look at a picture of the Crab Nebula, synthesized from visible-light Hubble photographs and Chandra X-ray images, and say "Oh my God?" Who can look at that and not experience a sense of wonder?

Who can watch people assembling the greatest engineering project in the history of mankind—the International Space Station—and not wonder at the ability of people to conceive and to execute that project? And it also addresses our sense of monument building, of leaving something behind for future generations. Not for nothing, 31 years after its opening, is the National Air and Space Museum still the most heavily visited museum in Washington, D.C. And what do people come to see? They come to see early airplanes and Apollo spacecraft.

Of course the space program also addresses the Acceptable Reasons I've mentioned. In the end, this is imperative. Societies will not succeed in the long run if they place their resources and their efforts in enterprises that, for whatever reason, don't provide concrete value to that society.

But my point earlier is that if things are done for the Real Reasons that motivate humans, they also serve the Acceptable Reasons. In that sense, in the practical sense, space really is about spin-offs, as many have argued. But it's not about spin-offs like Teflon, Tang and Velcro as the public is so often told—and which in fact did not come from the space program. And it's not about spin-offs in the form of better heart monitors or cheaper prices for liquid oxygen for hospitals. Yes, you get those things and many more; and they are real benefits. But that's not the right level on which to view the matter. The real spin-offs are at a higher level. We need to look at a broader landscape.

What is the economic value to a society of upgrading the precision to which the entire industrial base of that society works? Anyone who wants to put together space artifacts, who wants to bid on a competition for space artifacts, who wants to be a subcontractor or supplier or who even wants to supply nuts, bolts and screws to the space industry, must work to a higher level of precision than human beings had to do before the space industry came along. And that fact absolutely resonates throughout our entire industrial base. What is the value of that? I can't calculate it, but I know it's there.

What is the scientific value of discovering the origins of our universe? Or of discovering that literally 95 percent of the universe consists of dark energy or dark matter, terms for things that we as yet know nothing about? But they make up 95 percent of our universe. Is it even conceivable that one day we won't learn to harness them? As cavemen learned to harness fire, as people two centuries ago learned to harness electricity, we will learn to harness these new things. It was just a few years ago that we discovered them, and we would not have done so without the space program. What is the value of knowledge like that? I cannot begin to guess. A thousand years from now there will be human beings who don't have to guess; they will know; and they will know we gave this to them.

Let's think for a moment about national security. What is the value to the United States of being involved in enterprises that lift up human hearts everywhere when we do them? What is the value to the United States of being engaged in such projects, doing the kinds of things that other people want to do with us, as partners? What is the value to the United States of being a leader in such efforts, in projects in which every nation capable of doing so wants to take part? I would submit that the highest possible form of national security, well above having better guns and bombs than everyone else, well above being so strong that no one wants to fight with us, is the security which comes from being a nation which does the kinds of things that make others want to work with us to do them. What security could we ever ask that would be better than that, and would give more of it to us, than the space program?

What do you have to do? How do you have to behave, to do space projects? You have to value hard work. You have to live by excellence, or die from the lack of it. You have to understand and practice both leadership and followership; and both are important. You have to build partnerships; leaders need partners and allies, as well as followers. You have to be willing to defer gratification, to spend years doing what we do, and then stand back and see if it works. We learn how to leave a legacy, because we work on things that not all of us will live to see—and we know it. And we learn about accepting the challenge of the unknown, where we might fail, and to do so not without fear or apprehension, but to master it, to control it and to go anyway.

These are lessons that we all need to learn, and they are lessons the space business teaches us. And I would submit that our country is a better place for those who have learned those lessons.

These are the values that the space program brings. This is why it must be supported. And this is why, although we don't acknowledge it, we don't admit it and most of us don't understand it. This is why if we didn't have a space program, we Americans would feel less than ourselves. We can never allow that to happen.

Why Explore Space?

Michael Griffin
Administrator
National Aeronautics and Space Administration

Op-ed posted on *www.nasa.gov* on January 18, 2007, and distributed nationwide on the McClatchy/Tribune Wire Service for publication in local and regional news publications.

As NASA resumes flights of the space shuttle to finish building the International Space Station, many are questioning whether the project—the most complex construction feat ever undertaken—is worth the risk and expense.

I have been asked, and asked myself, this question many times during my career, particularly when the United States lacked a plan to go beyond the space station to other destinations in the solar system.

The issue was addressed eloquently in the report of the *Columbia* Accident Investigation Board, which examined the 2003 loss of the shuttle and its crew. That report pointed out that for the foreseeable future, space travel is going to be expensive, difficult and dangerous. But for the United States, it is strategic. It is part of what makes us a great nation. And the report declared that if we are going to send humans into space, the goals ought to be worthy of the cost, the risk and the difficulty. A human spaceflight program with no plan to send people anywhere beyond the orbiting space station certainly did not meet that standard.

President Bush responded to the *Columbia* report. The administration looked at where we had been in space and concluded that we needed to do more, to go further. The result was the Vision for Space Exploration, announced nearly 3 years ago, which commits the United States to using the shuttle to complete the space station, then retiring the shuttle and building a new generation of spacecraft to venture out into the solar system. Congress has ratified that position with an overwhelming bipartisan majority, making the Vision for Space Exploration the law of the land.

Today, NASA is moving forward with a new focus for the manned space program: to go out beyond Earth orbit for purposes of human exploration and scientific discovery. And the space station is now a stepping stone on the way, rather than being the end of the line.

On the space station, we will learn how to live and work in space. We will learn how to build hardware that can survive and function for the years required to make the round-trip voyage from Earth to Mars.

If humans are indeed going to go to Mars, if we're going to go beyond, we have to learn how to live on other planetary surfaces, to use what we find there and bend it to our will, just as the Pilgrims did when they came to what is now New England—where half of them died during that first frigid winter in 1620. There was a reason their celebration was called "Thanksgiving."

The Pilgrims were only a few thousand miles from home, and they were accomplished farmers and artisans. And yet, when they came to an unfamiliar land, they didn't know how to survive in its harsh environment. They didn't know what food would grow and what wouldn't. They didn't know what they could eat and what they couldn't.

The Pilgrims had to learn to survive in a strange new place across a vast ocean. If we are to become a spacefaring nation, the next generation of explorers is going to have to learn how to survive in other forbidding, faraway places across the vastness of space. The moon is a crucially important stepping stone along that path—an alien world, yet one that is only a three-day journey from Earth.

Using the space station and building an outpost on the moon to prepare for the trip to Mars are critical milestones in America's quest to become a truly spacefaring nation. I think that we should want that. I want that. I want it for the American people, for my grandchildren, for my great-grandchildren.

Throughout history, the great nations have been the ones at the forefront of the frontiers of their time. Britain became great in the 17th century through its exploration and mastery of the seas. America's greatness in the 20th century stemmed largely from its mastery of the air. For the next generations, the frontier will be space.

Other countries will explore the cosmos, whether the United States does or not. And those will be Earth's great nations in the years and centuries to come. I believe America should look to its future and consider what that future will look like if we choose not to be a spacefaring nation.

Space Exploration: Filling Up the Canvas

Michael D. Griffin
Administrator
National Aeronautics and Space Administration

NASA Langley Colloquium Series
Sigma Public Lecture Series
October 24, 2006

Thank you for inviting me to speak at this colloquium; I am truly honored. This lecture series dates back to 1971, with the inaugural address by Wernher von Braun. Many other luminaries from our industry have followed him, so I have big shoes to fill. I started my career in the aerospace business in that same year. Maybe there is a young person in this audience who will be giving the 70th anniversary lecture on the then-future of space exploration, 35 years from now.

This area of our country, Virginia, has given birth to many great leaders and explorers, whose ideas for our nation's future speak to us across the generations as we carry out the great task before us in space exploration. No event in our history more aptly conveys those ideas and lessons than does the Lewis and Clark expedition, which concluded exactly 200 years ago. When discussing this great expedition, led by Virginians Meriwether Lewis and William Clark, fellow Virginian Thomas Jefferson wrote, "The work we are now doing is, I trust, done for posterity in such a way that they need not repeat it. We shall delineate with correctness the great arteries of this country. Those who come after us will fill up the canvas we began."

Today, we are endeavoring to "fill up the canvas" of our solar system in such a way that our work is done for posterity as well. When President Bush laid out the canvas for NASA with the Vision for Space Exploration in 2004, he evoked the Lewis and Clark expedition, saying: "Two centuries ago, Meriwether Lewis and William Clark left St. Louis to explore the new lands acquired in the Louisiana Purchase. They made that journey in the spirit of discovery, to learn the potential of vast new territory, and to chart a way for others to follow. America has ventured forth into space for the same reasons. We have undertaken space travel because the desire to explore and understand is part of our character."

When President Bush set this new course for America's space program, the White House issued a supporting document explaining why. Quoting from that policy, "The fundamental goal of this vision is to advance U.S. scientific, economic and security interests through a robust space exploration program." I believe that this is exactly right, and that the benefits to be derived in these respects from such a program were exactly the same ones that Jefferson expected to derive from the Louisiana Purchase and from the expedition he sent out to begin its assessment.

Security in Jefferson's time meant establishing the primacy of the infant United States across the breadth of the North American continent, in an era when numerous competitors for this primacy existed. Today's nation, stretching "from sea to shining sea," was the vision of farseeing men like our third president, but few others.

The Lewis and Clark expedition paved the way for future adventures by the new nation in what eventually became the American West. Indeed, would it even have become the "American" West without the staggering success of this first great westward trek? At the time of their expedition, Spain, France, England and Russia had interests and a presence in what is today the Western United States. This is a message that speaks to us today, across two centuries of time, as we contemplate the future of humans in the solar system.

In our time, while we certainly recognize that the United States will be only one nation among many on the space frontier, we have learned that "security" can involve much broader concerns than competition among nation-states. The Chairman of the NASA Advisory Council, Harrison Schmitt, geologist, Apollo 17 astronaut, and former United States Senator and Stephen Hawking, cosmologist and Lucasian professor of mathematics at the University of Cambridge, have both pointed out this fundamental truth: The history of life on Earth is the history of extinction events. There is evidence, now, for some five major extinction events in the history of the planet. The last of these occurred approximately 65 million years ago, at the end of the Cretaceous Period, when the dinosaurs that dominated Earth for over 160 million years suffered a catastrophic extinction over a relatively short period. It is believed that this event was induced by an asteroid of some 6–15 kilometers in diameter which struck Earth in the Gulf of Mexico, triggering tsunamis, tectonic shifts and radically changing Earth's atmosphere.

The brief history of humans is next to nothing compared to the history of other life on Earth, and even less so compared to the age of our solar system or of the universe. Our species hasn't been around long enough to have experienced a cataclysmic extinction event. But they will occur again, whether we are ready for them or not. So, in the end, human expansion into our solar system is fundamentally about the survival of the species, about ensuring better odds for our survival through the promulgation of our species. There is no more fundamental measure of "security."

But security is not the only reason to explore. History shows clearly that there is an economic benefit to be derived from exploring new territories. Jefferson understood this. When he proposed the Lewis and Clark expedition in a secret message to the Congress, he said: "While other civilized nations have encountered great expense to enlarge the boundaries of knowledge by undertaking voyages of discovery, and for other literary purposes, in various parts and directions, our nation seems to owe to the same object, as well as to its own interests, to explore this, the only line of easy communication across the continent, and so directly traversing our own part of it. The interests of commerce place the principal object within the constitutional powers and care of Congress, and that it should incidentally advance the geographical knowledge of our own continent, cannot be but an additional gratification."

Likewise, the Vision for Space Exploration recognizes the economic benefits to be derived from space exploration. As the President's Science Advisor Jack Marburger stated in a speech earlier this year, "Questions about the vision boil down to whether we want to incorporate the solar system in our economic sphere, or not. Our national policy, declared by President Bush and endorsed by Congress last December in the NASA Authorization Act, affirmatively answers that question: The fundamental goal of this Vision is to advance U.S. scientific, security and economic interests through a robust space exploration program."

In this vein, the U.S. segment of the International Space Station has been designated a national laboratory, open for commercial manufacturing and advances in materials sciences due to its unique microgravity environment. To that end, I commend the Langley Research Center's material scientists for recently retrieving 200 specimens from the Materials International Space Station Experiment (MISSE) "suitcase" with the space shuttle. Working with the Naval Research Laboratory and others, these experiments may lead to more advanced solar arrays, and help researchers make materials and coatings that last longer on Earth.

On the moon, there are resources to be mined, including hydrogen, oxygen and maybe one day helium-3, which could be of special benefit in establishing a permanent lunar presence. An armada of satellites from the United States, India, China and Japan is set to map the moon's geography and resources over the next several years in anticipation of future human exploration and, potentially, lunar settlements. Of particular interest are the moon's polar regions, where some locations enjoy near-permanent sunlight while others, only a few kilometers apart, are permanently shadowed. The former are obviously of tremendous benefit in establishing a lunar base, because of the ability to generate nearly continuous solar power. And in the shadowed regions, it is possible that water ice deposited by comet impact might be found preserved from evaporation by the sun's heat. If such ice exists, it would be a boon for a future lunar base, enabling the occupants to "live off the land" more easily than carrying all provisions with them from Earth.

But this rosy prospect is, at present, still conjecture. While there is some evidence to support it, there remains considerable debate among lunar scientists as to whether such conditions truly exist at the poles, and, if so, how much ice is there. We won't know until we conduct a better survey. This debate among lunar scientists is not unlike the debate about the unknown geography of our own North American continent at the time of the Lewis and Clark expedition.

The next robotic lunar missions will test our assumptions and challenge our beliefs. But one assumption that I know will be justified is that the moon, the near-Earth asteroids and the rest of the solar system contain the resources that will take mankind to the next level of civilization and prosperity. I don't know when it will occur or who will do it, but it will happen. I hope that it will be soon, and that we will be the agents of this great endeavor.

Jefferson was the most scientifically literate president our nation has had, and he fully understood that his bold expedition would, almost automatically, open a new realm of scientific discovery. Jefferson's instructions to Meriwether Lewis in June of 1803 read like a NASA requirements document today: "... explore the Missouri river, and such principal stream of it as by its course and communication with the waters of the Pacific Ocean whether the Columbia, Oregon, Colorado or any other river may offer the most direct and practicable water communication across this continent for the purposes of commerce." Jefferson's additional requirements for the Lewis and Clark mission: "[Y]ou will take careful observations of latitude and longitude at all remarkable points on the river. ... Other objects of worthy notice will be the soil and face of the [territory] its growth and vegetable productions, ... the animals of the [territory], ...

the mineral productions of every kind; but more particularly metals: limestone, pit-coal and salt-petre; salines and mineral waters … volcanic appearances and climate." This was advanced scientific inquiry for that day.

For NASA, exploration is about the expansion of human and robotic activity beyond Earth. This sets the stage for scientific opportunities which we are just now beginning to consider. Soon, we will begin to add to our civilization's body of knowledge concerning the real estate values in cislunar space, and we will conduct scientific experiments along the way, much in the fashion that Meriwether Lewis and William Clark gathered specimens, made careful observations in their journals and drew detailed maps of the American West 200 years ago.

Jefferson's plans were comprehensive, yet flexible. The hoped-for water route to the Pacific did not, in fact, exist. Yet, because the expedition did not have a single overriding goal, it was an enormous success. This should also be our strategy when making plans to explore the moon and Mars. We should expect to be surprised, and we should adjust our exploration plans as we learn more about the lay of the land before us, its resources and environmental conditions.

If we are able to live and work on the moon, we will not only use its resources for our survival and economic benefit. We will think of ways to exploit its unique vantage point and environment to further our scientific goals. Back in 1990, the National Academy of Sciences studied the suitability for using the moon as a stable platform, without an atmosphere and having predictable heating and lighting, for astronomical observatories, especially interferometers. Going into the next decadal study for astronomy and astrophysics, the Academy should consider how we can better leverage the exploration architecture to further scientific pursuits "and other literary purposes" as Jefferson would say, so that we can plan our expeditions appropriately.

So, what is our approach to achieving the goals of which I have spoken here?

Our nation's Vision for Space Exploration honors our previous commitment to the space station and at the same time commits us to bold new journeys to the moon, Mars and eventually the rest of the solar system, to learn the potential of this vast new territory and chart a way for others to follow. With our Russian, European, Japanese and Canadian partners, the United States is completing the assembly of the space station. We will then retire the space shuttle in 2010. Meanwhile, we are beginning to build new space ships and rockets to carry astronauts and, one day, future settlers outward from low Earth orbit.

The scientists and engineers of Langley Research Center are integral to turning this vision into reality. Experts in structures, materials and other disciplines in the aerospace sciences, along with the NASA Engineering and Safety Center (NESC) that is hosted at Langley, helped return the space shuttle to flight after the *Columbia* accident. Their work has been instrumental in understanding the physics behind foam loss on the external tank, and its effects on the shuttle thermal protection system. Langley engineers worked on the computational fluid dynamics (CFD) analysis to support the removal of the shuttle External Tank Protuberance Air Load (PAL) ramps. All of this was absolutely crucial to the future of our agency; absolutely nothing good can happen at NASA unless we can fly the shuttle with confidence that we have fixed the problems that brought down *Columbia*.

Looking to the future, the NESC organized a "smart buyer" team across the agency earlier this year to conduct an "in house" design of our Orion Crew Exploration Vehicle, so that managers and engineers could better evaluate industry designs and sharpen the systems engineering and integration skills needed to manage this major undertaking. And a project team hosted at Langley is managing the Orion Launch Abort System, which we hope to test beginning in 2008. Others are working on the Orion landing system vertical drop tests with the half-scale crew module. And yet others have conducted wind tunnel tests of the Ares 1 Crew Launch Vehicle to characterize the launch vehicle stack.

This stuff is rocket science! As an engineer myself, I fully appreciate the challenges before us, and frankly, we should all recognize that the development of the arts and sciences of spaceflight is quite simply the most technically challenging thing our nation, or any nation, does. Meriwether Lewis' perspective on the challenges ahead of him on July 4, 1805 speaks to many of us in NASA today: "We all believe that we are now about to enter on the most perilous and difficult part of our voyage, yet I see no one repining; all appear ready to meet those difficulties which wait us with resolution and becoming fortitude."

We must also recognize the dangers involved. Virginian David Brown, no less an explorer than any on the Lewis and Clark expedition, died with his fellow crewmates on Space Shuttle *Columbia*. A graduate of William and Mary and Eastern Virginia Medical School, David once said that even in the case of a possible catastrophe for his upcoming mission, "The program will go on. It must go on."

Thomas Jefferson was equally cognizant of the perils awaiting the Lewis and Clark expedition. In his letter of instructions to Meriwether Lewis, Jefferson wrote: "As it is impossible for us to foresee in what manner you will be received by the native people, whether with hospitality or hostility, so it is impossible to prescribe the exact degree of perseverance with which you are to pursue your journey. We value too much the lives of our citizens to offer them to probable destruction. … To your own discretion therefore must be left the degree of danger you risk, and the point at which you should decline to continue, only saying we wish you to err on the side of your safety, and to bring back your party safe even if it be with less information that you will have acquired to that point." For the record, Lewis and Clark succeeded admirably in this matter, while still achieving the larger goals of their venture. Only one man was lost on the expedition, from what was believed to have been a burst appendix, an ailment which could not have been treated in that era in any case.

Some days our journey into space must appear altogether boring to the casual observers, the pundits or the "chattering class," as they're sometimes called in Washington, who are not steeped in the trials and tribulations of great challenges. The critics will never appreciate the hard but tedious work, and the sheer joy, that goes with the successful accomplishment of every space shuttle flight, or a record-breaking hypersonic flight like that of the X-43A, or the development of a new satellite capability, like the Cloud-Aerosol Lidar and Infrared Pathfinder Satellite Observation (CALIPSO) lidar instrument managed at the Langley Research Center.

This also is not new. The daily entries in the Lewis and Clark journals are sometimes filled with "nothing to report" as well, as the men of the Corps of Discovery and their female Indian interpreter Sacagawea endured the daily rains in Oregon. However, historians have noted the three words in Meriwether Lewis's journal that are often-repeated, and are the most important in understanding the character of those making such a journey: "We proceeded on." Lewis repeats this phrase in his journal on many days, after attacks by native Indians and grizzly bears, after seeing great bison stampedes, after capturing a prairie dog, after backbreaking portages with their canoes and after gazing upon the daunting mountain ranges which they had to traverse to reach the West Coast of America. Indeed, "we proceeded on" evokes the sense of determination that David Brown expressed about our basic human need to explore.

The Vision for Space Exploration carries on the tradition of exploration embodied by two Virginians of whom we have spoken tonight, Meriwether Lewis and William Clark, 200 years ago. They carried out their mission for very similar reasons that we carry out our mission today—national security, economic gain and scientific discovery. While space exploration is certainly fraught with difficulty and peril, we can at the same time both appreciate those risks and yet believe that for the same reasons as existed 200 years ago, this is truly the most rewarding endeavor our nation will pursue in the 21st century.

When Lewis and Clark and other members of the Corps of Discovery returned and, subsequently, were feted for their accomplishments in Washington, one senator remarked that they appeared "as if they had returned from the moon." How apt. The Lewis and Clark expedition embodied the pioneering spirit which is characteristic of our nation, the spirit which led us forward to the Apollo 11 lunar landing by Neil Armstrong and Buzz Aldrin. Lewis and Clark made "one giant leap for mankind," right along with Armstrong and Aldrin. New leaps will soon follow.

The next steps in returning to the moon and moving onward to Mars, the near-Earth asteroids and beyond are crucial in deciding the course of future space exploration. We must understand that these steps are incremental, cumulative, and incredibly powerful in their ultimate effect. As President Bush pointed out when announcing the Vision for Space Exploration, "We will make steady progress—one mission, one voyage, one landing at a time." Further, we must understand that there is no turning back. In the words of David Brown: "It must go on."

Allow me to end with the thoughts of Meriwether Lewis on the day he turned 32 years old. Lewis was on one of the greatest journeys of his time, of any time, yet he did not realize its significance while he was doing it. Instead, he was consumed with the great mission before him. Jefferson once opined that Lewis suffered from a certain melancholy when it came to his work. Meriwether Lewis wrote the following passage of enlightenment in his journal on August 18, 1805: "This day I completed my thirty first year, and conceived that I had in all human probability now existed about half the period which I am to remain in this Sublunary world. I reflected that I had as yet done but little, very little, indeed, to further the happiness of the human race, or to advance the information of the succeeding generation. I viewed with regret the many hours I have spent in indolence, and now sorely feel the want of that information which those hours would have given me had they been judiciously expended. But since they are past and cannot be recalled, I dash from me the gloomy thought, and resolved in future, to redouble my exertions and at least endeavor to promote those two primary objects of human existence, by giving them the aid of that portion of talents which nature and fortune have bestowed on me: or in future, to live for mankind, as I have heretofore lived for myself."

So, in conclusion, I really do hope that there is a young person in the audience today who, many years from now, will continue the tradition of this lecture series by telling us how she helped to fill up the canvas, breaking the confines of this "sublunary" world.

Thank you.

Continuing the Voyage: The Spirit of Endeavour

Michael D. Griffin
Administrator

National Aeronautics and Space Administration
United States of America

Remarks to
The Royal Society of the United Kingdom
December 1, 2006

I am truly honored to have been invited to speak before this august group. The Royal Society has a long and distinguished history of supporting explorers and scientists; indeed, for centuries the Royal Society was the embodiment of science in western civilization. Yesterday, I had the honor of making a very special presentation of the Society's prestigious Copley Medal to Professor Stephen Hawking of Cambridge University. This medal, the world's oldest award for scientific achievement, flew on board Space Shuttle *Discovery* on the STS-121 mission last summer, part of the personal effects of British-born astronaut Piers Sellers, as a gesture in honor of Professor Hawking.

I hope that next week Space Shuttle *Discovery* will roar back into space with the STS-116 mission to continue the assembly of the International Space Station.[1] When it does fly, another British-born astronaut, Cambridge graduate Nick Patrick, will be aboard. The upcoming shuttle missions to finish the space station are among the most difficult and complex ever undertaken. On this mission alone, the crew will add another segment to the space station truss, the backbone of the configuration. They will re-configure the electrical power system to incorporate and use the new solar arrays brought up on the last flight, and they will fill and activate the ammonia cooling system for two of the truss segments. The exploration and development of the space frontier is, truly, the most technically challenging endeavor of our generation and many to follow.

---

[1] STS-116 launched on December 9, 2006 and returned after a successful mission on December 22, 2006.

But in carrying it forward, we are building on the heroic exploits of our forbears in their own missions of human exploration and scientific discovery.

I have no greater personal hero than former British Prime Minister Winston Churchill, a man whose distinguished paternal ancestors served Britain for centuries, yet who was born of an American mother. Throughout my life I have admired Churchill's famed incisive wit, his stunning oratorical skills, his steadfastness in support of that in which he believed, and above all the unbreakable rock of his courage. During the darkest days of World War II, as he sought to bolster his countrymen and those everywhere who fought for freedom, Churchill exclaimed, "We have not journeyed all this way across the centuries, across the oceans, across the mountains, across the prairies, because we are made of sugar candy!"

Nor will we journey into space, and make the solar system our own, by being made of sugar candy. The United Kingdom is a truly great nation, a nation whose people and culture have spread across the globe, and whose language has become the world's most common second language, due in no small measure to the support of Royal Society for its nation's explorers over the centuries. As we consider the migration of human beings out into space, first to the moon and Mars and then eventually beyond, I think it is interesting to look back and consider the migration of the human species and its languages and cultures to all of the continents of this planet. Our forebears have left us a history filled with lessons for the future. Some of these lessons are in evidence in my own land. America's origins do not begin on a specific date, nor do they involve any one particular group of people. Many of us in America are the descendants of pioneers from Spain, Portugal, Holland, Scotland, England, Ireland, France, Germany, Italy and many other countries, who emigrated over many generations and settled in what became the United States, in search of new riches, new freedoms and new beginnings. The several peoples of the British Isles were not even the first of these many groups, but they were in the end the boldest and most persistent. And so, over many generations, the primary language of the United States came to be English, and our dominant cultural traditions are derived from Great Britain. Thus, while we Americans are quite a mixed bag, in many respects we are your cultural, political and quite often genetic descendents. One of my own great-grandfathers emigrated from Scotland, another was Irish, a great-grandmother was a Hobbs, and my surname indicates the presence of a Welshman somewhere on the family tree! So my hope is that the English language will not only remain in common usage around the world, but it will spread throughout the solar system over the course of the next century as modern-day explorers like NASA astronauts Piers Sellers, Nick Patrick and others carry their British heritage with them into space. That is truly a lasting legacy for a great people.

Now, I am not a historian, but I am mindful of the lessons of history and how we can apply them to the challenges that NASA faces today in space exploration and scientific discovery. The Royal Society was one of the primary benefactors to many great maritime explorers of the 17th and 18th centuries. Today, in many respects, NASA is carrying on the same tradition of combining exploration and scientific discovery that the Royal Society initiated centuries earlier. Indeed, NASA is so beholden to the Royal Society and its traditions that two space shuttles, *Discovery* and *Endeavour*, were named in honor of sailing ships used by Captain James Cook, one of Britain's, and history's, greatest explorers. And when astronauts Dave Scott and Jim Irwin voyaged to the mountains of the moon on the Apollo 15 mission, their command ship was also named *Endeavour*.

So in drafting my speech for today, I thought it might be insightful to consider the connection between certain lessons from Captain Cook's initial South Pacific voyage on His Majesty's Bark *Endeavour*, to the work NASA is carrying out today in exploring the planets, moons, asteroids and comets of the solar system. Cook's first expedition to the South Pacific in 1768 was funded jointly by the Royal Society and the British Admiralty. The primary purpose of the voyage was to obtain astronomical observations of the transit of the planet Venus, as seen from Earth, across the disk of the sun on Saturday, June 3, 1769. The secondary intent of the voyage was to search the South Pacific for signs of a southern continent, Terra Australis, which had been conjectured to exist by members of the Society. He was only 39 at the time and, from my present vantage point, clearly far too young to be entrusted with such major responsibilities. What in heaven's name were his superiors thinking?

If I were speaking to the Royal Astronomical Society, everyone present would understand why the Royal Society was interested in having Captain Cook and a team of scientists observe the 1769 transit of Venus, and I could save a few words. But for tonight, we should probably note that prior to the invention of radar and, later, the capability to send spacecraft to other planets, it was extremely difficult to determine the actual size of the solar system. But, using a method first proposed by Sir Edmund Halley, himself an early and renowned member of the Royal Society, it is possible to use observations of a Venus transit to calculate the distance from Earth to the sun, the "astronomical unit," or A.U., and with that to determine the scale of the solar system.

Venus transits are one of the rarest predictable celestial events, typically lasting for only a few hours at a time. Between the 12th and 39th centuries they occur in a 243-year cycle, with appearances in pairs 8.5 years apart, separated by gaps of over a century. The most recent Venus transit occurred in 2004; the next one will be in 2012; and after that, we will need to wait until 2117 for another. I observed the 2004 event personally. And it may not be too much to hope for that I will see the 2012 transit, but I expect I shall miss the one after that.

The practical difficulties of making such observations were substantial in the 18th century. It is necessary to measure, quite accurately, the entry and exit times at which Venus crosses the limb of the sun, as seen from widely separated points on Earth. So, to begin, it was necessary to meet the fundamental challenge of merely traveling to various distant points on the globe, including the South Seas. This was an enormously difficult prospect in its day. Considered objectively, one must say that it was fully as difficult, dangerous and time consuming for Cook and *Endeavour* to reach the South Pacific as it will be for the first voyagers to reach Mars. Indeed, my personal opinion is that it was far more so, given the technology of those times. I believe that this is an important perspective for those who believe, somehow, that the exploration of space is uniquely difficult in comparison with the exploration of Earth by Europeans. And so this is the first lesson. Are we to quail before multiyear voyages to uncertain destinations, when our ancestors did not?

Among the many challenges of long sea voyages in the 18th century was, first, the basic task of determining one's location! Accurate and consistent timekeeping at widely separated points, equivalent to knowing one's longitude, was still a major challenge in 1769, and nearly impossible to do while aboard ship. Today, we take for granted Global Positioning System (GPS) navigators that receive precise timing signals from satellites with atomic clocks in orbit around Earth. Back in 1769, Captain Cook did not even have the benefit of an accurate chronometer. It was not until his second voyage to the South Pacific a few years later that Cook carried with him the K1, a copy of Harrison's H4, the clock that won him the famed Longitude Prize in 1773. Indeed, modern-day navigators and timekeepers, using GPS, are forever indebted to John Harrison and his famous clocks, some of which can be seen today at the Greenwich Observatory.

Captain Cook learned the value of accurate navigation and precise timing in the late 18th century; and the ability to carry out the primary purpose of his voyage was only barely possible with the technologies available to his expedition at that time. He didn't even have a good map; indeed, his job was in part to help make them. Modern-day explorers and scientists also know the value of a good map and accurate GPS measurements. Data from NASA's LANDSAT satellites provides the backdrop for maps provided by Google Earth and others. We will need a similar navigation infrastructure on the moon for future explorers and scientists. Scheduled for launch in 2008, NASA's Lunar Reconnaissance Orbiter, with its laser altimeter and other instruments, will provide an accurate global map of the moon for future explorers. We're still formulating our plans for providing communication and navigation for future explorers on the moon, but I can foresee NASA collaborating with other spacefaring nations like the United Kingdom in providing such infrastructure.

Such collaborations again have a long and honorable history. In another interesting parallel to space exploration today, the effort to observe the 1769 transit of Venus was an early example of international scientific collaboration. Cook's expedition to the South Seas and his sighting of the Venus transit from Tahiti was but one of many similar efforts, with scientists and explorers from Britain, Austria, France and other countries traveling to Siberia, Norway, Madagascar and the southern tip of Africa. Catherine the Great of Russia even invited astronomers to observe the transit of Venus from her observatory in Saint Petersburg.

Similarly, and as of today, NASA has 58 ongoing space and Earth science missions, and over half of these missions have some form of international participation. Two-thirds of all NASA missions currently under development incorporate international partners. And of course, NASA's premier human spaceflight program, the development of the space station, is an effort involving some 15 nations.

Like the collaboration for the Venus transit, NASA's partnerships in space exploration and scientific discovery take many forms, with various levels of contribution. For example, we are contributing two payloads to India's Chandrayaan-1 mission to the moon, planned to be launched in 2008. For the Cloud-Aerosol Lidar and Infrared Pathfinder Satellite Observation (CALIPSO) mission launched earlier this year, NASA developed a LIDAR payload; the French Space Agency (CNES) integrated that payload to its satellite bus and NASA launched it. The British National Space Center provided a high-resolution atmospheric sounding payload to the Aura Earth science mission launched two years ago. In the next decade, the European Space Agency will launch the James Webb Space Telescope aboard an Ariane V rocket.

One of the more unusual aspects of Cook's first expedition is that few members of the crew of *Endeavour* contracted scurvy, a disease now known to result from a lack of Vitamin C. By the mid-1700s, it was widely known that a poor diet caused scurvy, but what specifically caused it was not known. Captain Cook led by example and motivated his sailors into eating Vitamin C-rich (but not very tasty) sauerkraut by being the first to eat it during his meals.

While today's space station crewmembers don't get scurvy, they face other medical issues. For example, muscle and bone density loss due to a lack of tension on the human body in zero-G is well known. While there is a significant degree of variability between station crewmembers in the amount of bone loss, the average density lost in the spine and hip areas is about one percent per month. This rate of bone loss for astronauts is 10 times worse than for those who suffer from osteoporosis here on Earth. Thus, like sailors of the 18th century, our astronauts on the space station or in future missions to Mars face significant medical hazards in the form of bone fractures and kidney stones that could jeopardize their health and their mission. The equivalent of sauerkraut for modern-day astronauts is the unpleasant but necessary nutrition and exercise regimen to create muscle tension and mitigate bone loss. But these are stopgaps, incomplete and unsatisfactory at best. Whatever therapy is finally developed to control bone loss in astronauts will have application to sufferers of osteoporosis everywhere. Astronauts already have conducted clinical trials for new osteoporosis drugs onboard the space station. We have much to learn, and in learning we will create knowledge that can help people everywhere.

For exploration beyond low Earth orbit, radiation is another hazard to be dealt with. (At this point, I will again quote Churchill, who famously said that the grammatical prohibition against ending a sentence with a preposition is "an inconvenience up with which I shall not put.") The Earth's atmosphere, and especially its magnetic field, shields us from nearly all of the effects of solar flares and galactic cosmic radiation, even to the extent of providing substantial protection for low-orbiting astronauts. Despite this shielding effect, periodic and highly intense solar storms wreak havoc with power grids on Earth and satellites in high orbit. On several occasions, space station astronauts have hunkered down in heavily shielded areas of the station when solar flares or coronal mass ejections were predicted to be heading toward Earth. And as we venture farther away from Earth, the need to protect them from this energetic particle radiation becomes more critical. For example, back in August 1972, between the Apollo 16 and 17 missions, a powerful solar flare occurred that would have seriously endangered our astronauts if they had been en route to the moon or on the lunar surface at that time. An even larger sequence of solar events occurred in the fall of 2003. We must provide our astronauts with both warning systems and effective safeguards.

Looking through the eyes of multiple spacecraft over the past five decades, we have seen that in truth the planets of the solar system are embedded in the heliosphere, the exotic outer atmosphere of the sun, emanating from and shaped by its intense magnetic field. This heliosphere is analogous, in many ways, to the winds and currents of the Pacific Ocean that propelled Captain Cook's *Endeavour*, but which also endangered the vessel and crew during periodic storms. Thus, we must build our spaceships in ways that shield the astronauts and instruments inside; and we must provide timely warnings and predictions of "solar storms" just as we do now with weather forecasts here on Earth. Space weather monitoring and forecasts need to be extended beyond low Earth orbit to cislunar space and, eventually beyond, when we begin missions to Mars. This effort to safeguard planet Earth and our astronauts from solar storms must be an international endeavor, just like information from an international network of weather satellites and forecast centers today is shared around the world. We will benefit all mankind in the process by planning our heliophysics missions together. Earlier this month, NASA launched the Solar Terrestrial Relations Observatory (STEREO) mission, consisting of two satellites intended to provide 3-D images of the effects of coronal mass ejections and other solar activity on Earth's magnetosphere. The STEREO mission had several international partners, and will provide warnings that are useful around the world, but NASA cannot "go it alone." We must work together.

Scientific discoveries are sometimes elusive, but we must persevere. As things turned out, despite the best efforts of Captain Cook and his men and the other international collaborators in measuring the Venus transit, the separate measurements taken by the various scientific expeditions onboard the *Endeavour*, at Point Venus in Tahiti and other places around the globe varied greatly, and were, inevitably, inconclusive. It turned out to be very difficult to determine the precise limb crossing times for Venus against the solar background. A now notorious observational surprise, the so-called "black drop effect," smears the image of Venus precisely as it becomes tangential to the solar limb. In the end, none of the observations of the 1769 transit, whether from Cook's team or others, were very good. The astronomical unit would not be accurately determined until the 1880s, when American Astronomer Simon Newcomb published a value of 149.6 million km, using data from the four prior Venus transits.

Thus, we must recognize that while the mysteries of the universe may not elude Stephen Hawking, these mysteries do frustrate the rest of us who are mere mortals. We must be resolute in our convictions, and despite setbacks, we must recognize that progress through human exploration and scientific discovery is a goal worthy of the costs and risks of the enterprise. Again to echo Churchill, we are not made of sugar candy.

But the disappointment encountered by Cook's expedition yields another lesson for space mission planners today. Cook's voyage was not judged, nor intended to be judged, solely on its ability to obtain accurate measurements of the transit of Venus. Indeed, Cook is today remembered above all else for his discovery, the first by European voyagers, of the previously unknown continent of Australia. Who today would label Cook's first South Pacific voyage a failure because the measurements of the Venus transit were inconclusive? We must remember this and similar experiences when our future space missions encounter difficulties. And we must plan them so that they are not hostage to a single piece of good fortune as a measure of their overall success.

There is yet another lesson to be gained by looking back across the centuries at the voyages of *Endeavour*. Although Captain Cook no doubt considered his vessel to be the apotheosis of the shipbuilding arts as they were then known, we can see in retrospect that the *Endeavour* occupies but one point along a curve of ever increasing maritime capability, beginning for Europeans with the Viking longships and culminating in today's supertankers. We must realize that there will be a similar performance curve for space systems, and that we have not as yet advanced very far along it. We are at the dawn of the space age; metaphorically, we are sending out longships. We have a long, long way to go to get to supertankers. We must constantly question our assumptions as to how we build and operate our spacecraft and launch vehicles, because we have a lot yet left to learn. Each generation of ships must improve upon the last. For example, the space shuttle system, including the orbiter, solid rockets and external tank, requires almost 2 million labor-hours to prepare for launch. We have analyzed the processing labor required for other launch vehicles, both foreign and domestic, manned and unmanned. Using these other launch vehicle systems operating costs and labor-hours as a guide, I believe that NASA's next generation crew launch vehicle, the Ares I, should require an order of magnitude fewer labor hours to process than the space shuttle. The savings in launch vehicle operating costs can then be applied to future systems and bolder missions. With the retirement of the space shuttle in 2010, and development beginning for NASA's new Orion Crew Exploration Vehicle and Ares I Crew Launch Vehicle, we are beginning to lay the keel for the next generation of spaceships. Now is the time to question our operating assumptions and technical dogmas.

Reflecting upon his journeys, Captain Cook once said, "I had ambition not only to go farther than any man had ever been before, but as far as it was possible for a man to go." While some people might ridicule such bold ambitions, I think that it reflects the determination innate in all of us to push the limits of our technological capabilities and human faculties. However, I will express a certain lament to those sympathetic and like-minded members of the Royal Society who are concerned, with me, that our broader society today often seems to suffer from a lack of imagination in grasping the importance of the challenge before us.

One of the minor misfortunes of modern life in our major cities is that our night-time lighting has drowned out our view of the rich constellation of stars and planets in the night sky, and we find other idle pursuits, such as television, to occupy our time. Thus, we today do not look up nearly as often to marvel at the beauty and mystery of the night sky as did our ancestors, who imagined the stars to represent constellations of mythological beasts and legends, while the planets represented gods. I am happy that we have progressed beyond this. To me, the view of Hadley Rille from a camera mounted on the Apollo lunar rover is more exciting than imagining the moon to be the Huntress Diana. But, there is no question that we modern folk are less concerned with the heavens than were our ancestors.

But if we do the right things, maybe we can alter this perspective. The British Royal Astronomical Society recently released a report advocating the expansion of British involvement in human space exploration. I hope that report receives sober consideration in the policy circles of the United Kingdom, and I hope that I can count on you to be among the international partners who, with the United States, work to develop the first permanent lunar outpost in the next decade.

Last month I made the decision, the culmination of 18 months of work by NASA engineers and scientists, that we could effectively and safely conduct a space shuttle servicing mission to the Hubble Space Telescope to extend the life and capabilities of this great observatory. I have been struck by the tremendously positive response this decision has received, by the way that people from all over the world have been awed and inspired by Hubble pictures revealing a few of the secrets of our universe. Hubble provides glimpses into the universe that are far, far beyond the scale of the astronomical unit, the objective for Cook's first voyage to the South Pacific.

The view of our vast universe provided by Hubble uplifts us; it gives us a measure of hope. It was the same when the first man flew in space, and when the first man set foot upon the moon. We see a little of it each time a space shuttle crew returns from yet another mission in the sequence necessary to assemble the space station, quite possibly the greatest construction project in the history of mankind. We will see it again, at its peak, when the first astronaut places her boot on the surface of Mars. The human species was not crafted solely for safe places and prosaic times. We are, each of us, descended from people who left their homeland in search of what lay beyond. Today, what lies beyond is space. And so, quoting Professor Hawking: "To confine our attention to terrestrial matters would be to limit the human spirit."

I believe with all my heart that, with the exploration of space, we have embarked upon the boldest human adventure yet conceived. We are limited only by our imagination, ambition, ingenuity, persistence and leadership. But, "We have not journeyed all this way across the centuries, across the oceans, across the mountains, across the prairies, because we are made of sugar candy!"

Thank you.

Incorporating Space into Our Economic Sphere of Influence

Michael D. Griffin
Administrator
National Aeronautics and Space Administration

World Economic Forum
January 26, 2007

Good evening. Thank you for inviting me to speak tonight. It is not often that an aerospace engineer is invited to speak to an economic forum. However, I took a business degree along with my engineering and physics coursework; and I appreciate the economic impact that space has on our society, especially practical applications like communications, navigation, weather and remote sensing satellites as well as the economic, national security and scientific benefits. And this says nothing of the less-quantifiable benefits of intellectual inspiration.

Some of us gathered here tonight grew up during the Apollo era of the 1960s, NASA's apotheosis. We watched science fiction movies and television shows that made us believe that we—all of us and not simply a few astronauts—could become space travelers. Arthur C. Clarke's and Stanley Kubrik's masterpieces of science fiction *2001: A Space Odyssey* projected onto the screen of our collective human consciousness a future for us where, by now, hundreds of people would be living and working in space stations orbiting Earth, and outposts would exist on our moon. We would be journeying to other planets in our solar system, just as our European forbears came to America looking for new beginnings. This space age vision of our future proved illusory for our generation for two fundamental reasons: the limitations of our economic resources and the limitations of technology. Neil Armstrong's "giant leap for mankind" was not a journey that could be sustained without a more concerted investment of time, resources and energy than followed his seminal achievement on July 20, 1969.

But I believe that there are economic and technological reasons why we can now begin to afford and sustain this Vision for Space Exploration in a fashion where we "go-as-we-pay," and why the nations of the world making such investments of time, resources and energy will find that the benefits far outweigh the costs and risks involved. We have the technology and economic wherewithal to incorporate the benefits of space into our sphere of influence—to exploit the vantage point of space and the space environment, and we have the natural resources of the moon, Mars and near-Earth asteroids. Space exploration is not simply this century's greatest adventure; it is an imperative that, if not pursued with some concerted effort, will have catastrophic consequences for our society. I realize this is a bold statement, so allow me to explain.

On the day before he was assassinated in Dallas, President John F. Kennedy was in San Antonio, where he spoke about space exploration. He invoked Irish writer Frank O'Connor, who told the story of "how, as a boy, he and his friends would make their way across the countryside, and when they came to an orchard wall that seemed too high, and too doubtful to try, and too difficult to permit their voyage to continue, they took off their hats and tossed them over the wall—and then they had no choice but to follow them." The United States, the European Union, Russia, China, Japan, India and others have tossed our caps over the wall of space exploration.

In that same speech, President Kennedy recited several technical advances from NASA's space program, explaining that "our effort in space is not, as some have suggested, a competitor for the natural resources that we need to develop Earth. It is a working partner and a co-producer of these resources." And he finished this speech with the recognition of the costs and risks involved with space exploration: "We will climb this wall with safety and with speed—and we shall then explore the wonders on the other side."

Even an emotionless engineer can be moved by President Kennedy's poetic framing of the issues of space exploration, but since this is an economic forum, let me now turn to the "dismal science." When President Kennedy spoke those words in 1963, the Gross Domestic Product of the United States was approximately $2.8 trillion in FY2000 dollars. In 2005, it was approximately $11 trillion in those same FY2000 dollars—four times larger. In 1963, the U.S. federal government spent approximately $600 billion, again in FY2000 dollars, with NASA's allocation representing 2.3 percent of that amount. At the spending peak of the Apollo program, NASA represented 4.4 percent of the federal budget. Today, with a U.S. federal budget of almost $2.5 trillion, NASA's budget represents about 0.6 percent of that.

Clearly our economy has grown, our society has changed, and our priorities for government spending have changed since 1963. Thus, in the latter half of the 1960s and early 1970s, our nation's leadership decided that we should not sustain such a high percentage of investment in the space program. In these years, the priorities of the U.S. federal budget changed to accommodate the escalating costs of the war in Vietnam, defense spending for the Cold War and Great Society programs. Today, the costs of the Global War on Terrorism, Hurricane Katrina recovery, Social Security and Medicare/Medicaid dominate our federal government spending. The costs of our nation's entitlement programs alone are projected to double in the next 10 years, from more than $1 trillion per year today to more than $2 trillion per year, as baby boomers like me begin to retire. By comparison, NASA's budget of $16.2 billion for this year is somewhere in the realm of what engineers call rounding error, at 0.6 percent of all federal spending.

Because of the magnitude of these changes over the last four decades, it is important to view our nation's investment in our civil space and aeronautics research program from this larger economic perspective, because some critics have questioned the value proposition of even the current investment in NASA. I believe that we must recognize that the development of space is a strategic capability for our nations, and that we must bring the solar system into our economic sphere of influence. And equally, I believe that NASA must leverage the great economic engine of our nation and the world. Thus, the companies and countries that many of you represent can take advantage of the trails we plan to blaze as we explore space, just as we leverage the capabilities you create.

As a U.S. federal agency, NASA expects only inflationary growth in our annual budget. Thus, we have adopted a "go-as-we-pay" approach for space exploration, science missions and aeronautics research. Thus, the primary pacing item for new ventures is our nation's ability to afford such capabilities.

Over the next 3 years, our highest priority is to complete assembly of the International Space Station and honor our agreements to our Russian, European, Japanese and Canadian partners in this venture. It will not be easy. The space station is the world's greatest engineering project, akin to such feats as the Great Wall of China, the pyramids of Egypt, the Panama and Suez canals or the sea walls of Venice. Friends of mine who worked on the Apollo program have conveyed to me their belief that the construction of the space station is just as tough a job.

There are many critics of this space station, just as there were critics of President Kennedy who called the Apollo program a "moondoggle." But I believe that the greatest achievement of the space station partnership is the partnership itself, and that's a tough thing to criticize. For over 6 years, astronauts and cosmonauts have been living and working together aboard the space station. For the United States, the station is a national laboratory in space, where we will conduct research to make future exploration to other planets in our solar system possible. I hope this partnership will reap even greater dividends as we explore space together over many future generations. The unifying vision that forged this partnership during the 1990s, prompted by the Gore-Chernomyrdin Commission, is what we endeavor to carry forward today. Our partnership has endured some hardships along the way, not least of which was the *Columbia* accident. I hope and believe that those hardships have built stronger bonds between us.

With the proper goals in mind, I believe the benefits of space exploration far outweigh the risks. Among the most practical of these is our work with hurricane-monitoring satellites, aircraft and sensors that allow meteorologists to track such storms and predict their severity and impact. Many people today do not even realize that their weather forecasts rely on information from space assets.

Broader misconceptions exist. NASA spinoff technologies were never Tang, Teflon or space pens. But while we actually can cite tens of thousands of legitimate technology spinoffs, including medical devices, fuel cells, batteries and even cordless tools, I would like to discuss a more seminal point. I want people to realize the key areas where NASA's space endeavors have created entirely new industrial capabilities that improve our fundamental way of life.

For example, NASA is one of the major consumers of liquid hydrogen to fuel our space shuttle and other rocket engines. Liquid hydrogen is also used in the manufacturing of metals, glass, electronics and even foods. When you hear the term "hydrogenated fats" applied to baked goods like pastries and bread, it means that liquid hydrogen was one of the ingredients. NASA is such a large consumer of liquid hydrogen that after Hurricane Katrina, we returned several hundred thousand gallons to the nation's reserve and delayed several space shuttle rocket engine tests to alleviate a national shortage when our nation's liquid hydrogen production facilities and supply lines were disrupted. Likewise, we are a major consumer of liquid oxygen. Our huge demand market for these propellants sparked fundamental improvements in the production and handling of these volatile substances. Today, the ready availability of liquid oxygen allows firefighters, emergency response teams and nursing homes to carry on their backs or in suitcases portable, hand-carried oxygen tanks. In the 1960s, only select hospitals could supply oxygen in hazardous oxygen tents.

I am sure that many of you would agree with me that the greatest revolution in our productivity and way of life has been the development of the personal computer, Internet and various handheld communication devices. Thirty-five years ago, engineers like me used three pieces of wood and a piece of plastic that moved—the slide rule—to make calculations. Thirty years ago, 1,000 transistors could fit on a silicon chip; today, it's 100 million. The cost of such chips has dropped by a factor of 100,000. Few people know that the development of the first microprocessors was born of a competition between Fairchild and Intel in the 1960s to build components small enough to fit in NASA spacecraft. This straightforward NASA technical requirement spawned a whole new industry that grew in ways few, except perhaps Gordon Moore, could predict. Necessity is the mother of invention, and I believe that we are at our most creative when we embark on bold ventures like the space program.

So, with the economic growth and technology development we have seen since the 1960s, I believe that we are now entering a Renaissance period of space exploration where we can realize the vision that eluded us earlier. And as in the Renaissance, wealthy individuals will play a role in advancing the work of our architects, engineers and technicians. These will be entrepreneurs who have made their wealth in other endeavors—Jeff Bezos from Amazon, Bob Bigelow from Budget Suites, Richard Branson from Virgin and Elon Musk of Paypal fame are examples. These gentlemen and others have put their personal time, resources and energy behind the notion that many more people can have personal experience in space than do so today. It is one thing to view pictures of Earth from the vantage point of space, even on an IMAX screen, but it is another thing entirely to see it with one's own eyes. Many friends of mine have spoken of the epiphany they experienced from this.

But let me be clear. NASA's job is not to sponsor space travel for private citizens. That is for private industry. My hope is the reverse; that when the public can purchase rides into space, NASA can leverage this capability. Likewise, I hope that one day NASA can leverage the expertise of companies not unlike FedEx or UPS today, to meet our cargo needs for the space station and future lunar outposts. And one day, maybe, astronauts onboard our Orion crew exploration vehicle on their way to the moon and Mars can top off on liquid hydrogen from commercially available orbiting fuel stations.

In the process of building these new space capabilities, these entrepreneurs, along with NASA and other companies, are hiring more aerospace engineers. I believe that a key measure of a society's economic growth is the extent to which we are educating a technically literate people who can build the infrastructure to advance that society. It is deeply troubling to me when education statistics for the United States indicate there are more bachelor's degrees in psychology being awarded than engineering degrees. I am sure that even the economics majors here can appreciate my concern!

Again, NASA hopes to leverage, to the maximum extent possible, the capabilities that space entrepreneurs hope to create. A few years ago, when I was in the private sector working at In-Q-Tel, I helped fund a small software company seeking a better approach to visualizing satellite imagery. Over the years, that company grew into the backbone for Google Earth. Now, we hope to "spin-in" that capability to visualize imagery from other planets in our solar system, like the moon and Mars, using data from various NASA satellites and the Mars rovers. By invoking such commercial capabilities, NASA can leverage the funding of other investors to our mutual benefit.

In conclusion, I would like to leave you with a final thought as to what might happen if we do not explore space, if we do not follow the cap we tossed over the wall in the 1960s. Last month in the journal *Science*, researchers examining the primordial material returned by NASA's Stardust space probe found that some of that material could not have come from the Kuiper Belt in the outer reaches of our solar system, but instead could only have come from our sun's core. Some of that material was even older than our own sun.

The history of life on Earth is the history of extinction events, with evidence for some five major such events in the history of Earth. The last of these occurred approximately 65 million years ago, when the dinosaurs that dominated Earth for over 160 million years suffered a catastrophic extinction. It is believed that this event was caused by a giant asteroid , which struck Earth in the Gulf of Mexico, triggering tsunamis, tectonic shifts and radically changing Earth's climate.

The brief history of humans is next to nothing compared to the history of other life on Earth, and even less so compared to the age of our solar system or of the universe. Our species hasn't been around long enough to have experienced a cataclysmic extinction event. But they will occur, whether we are ready for them or not.

In the end, space exploration is fundamentally about the survival of the species, about ensuring better odds for our survival through the promulgation of the human species. But as we do it, we will also ensure the prosperity of our species in the economic sense, in a thousand ways. Some of these we can foresee, and some we cannot. Who could claim that he or she would have envisioned the Boeing 777 after seeing the first Wright Flyer? And yet one followed the other in the blink of an historical eye.

For this and many other economic and scientific reasons, we must explore what is on the other side of that wall, walk in the footprints of Neil Armstrong and make that next giant leap for mankind.

The Role of Space Exploration in the Global Economy

Michael D. Griffin
Administrator

National Aeronautics and Space Administration

NASA's 50th Anniversary Lecture Series
Washington, District of Columbia
September 17, 2007

I often talk about very large-scale themes with regard to space exploration and our science and aeronautics program; and today, my thrust is going to be to try to link some of that up to your daily life. We often think in terms of spinoffs. I have never been unduly fond of that term, but I do think that there is a very strong link, the strongest possible link between doing the hardest things that human beings do, which is flying in space, and how it benefits the rest of our economy and, indeed, our whole way of life. We are gathered here to kick off this lecture series commemorating NASA's upcoming 50th birthday, but we are celebrating more than what NASA has done and the benefits that have followed. We are also, when we do this, celebrating who we are and who we can be as an American people.

Fundamentally, NASA opens new frontiers and creates new opportunities, and because of that, we are a critical driver of innovation, but not in a way that just creates jobs. We create new markets and new possibilities for economic growth that didn't previously exist. We have taken at NASA to calling this the "space economy." It is an emerging economy, but a robust one even so, and it is an economy that is transforming lives here on Earth in ways that are not yet fully understood or appreciated. It is not an economy in space, or at least not yet, but space activities create products and markets that provide benefits right here on Earth, benefits that have arisen from our efforts to explore, understand and utilize the new medium.

In its last space report published in October 2006, the U.S. Space Foundation estimated the space economy at approximately $180 billion in 2005[1]. Over 60 percent of that figure derived from commercial goods and services. That is a stunning statistic for an economy that arises fundamentally the world over out of government programs. This growing economy affects just about every aspect of how we live, work and play. Other emerging new markets are just around the corner. The space economy enables satellite communications, including radio and television and telemedicine, point-to-point GPS navigation, weather and climate monitoring and a host of space-based national security assets. It also includes the nascent space tourism industry and the development of space logistics services that will transform space transportation into a viable commercial enterprise.

Fifty years ago, space was a far-off place. It was the stuff of science fiction. Today, it is pervasive in our lives, critical to a range of activities that create and provide value to human beings. It grew from NASA's roots in space exploration, and I would like to talk to you about that for just a few minutes. People all across our country, all across the world, find what we do exciting. They find it inspiring, and they find it so for many different reasons. Among them are the courage and competence of our astronauts, the dedication of the engineering teams that put them into space, the quest for knowledge that is realized by awe-inspiring pictures of distant galaxies or Martian craters from a robot perched on the rim, the challenge of the frontier, the final frontier, the frontier that begins anew on each planet and with each new discovery, the way we take on seemingly impossible tasks, tasks that challenge human skill and ingenuity to the utmost, like building a million-pound space station as a toehold on that final frontier.

What we do at NASA is quite simply larger than life. It is bolder than the boldest dreams, and we know it. So everyone knows and appreciates NASA,

---

[1] The 2008 report, released in April, estimated that the space economy had grown to $251 billion in 2007.

but to most people, what we do is literally out there. It is out of this world. The daily immediate connection between what we do and its impact on our lives is either unnoticed or taken for granted. In part, this is due to circumstances or, more properly, a change in circumstances because it wasn't always so. NASA was born and came of age during the cold war in an historical context that is difficult for many who were not there at the time to appreciate. It was a time when our very way of life was being called into question. The Soviet Union had declared that our democracy was too weak and too inefficient to compete with communism. After the successful launch of Sputnik, there were many people in our country and in the world who feared that they might be correct, and there were many others who were committed to proving them wrong.

The moon race was more than exploration for its own sake, and it was a lot more than an exercise in national pride. It was considered a real-live test of the viability of an open society, a vindication of the very concept of freedom. The American people admired NASA's expertise, our daring do or can-do attitude. These were a reflection of America itself. People marveled at our ability to meet John F. Kennedy's challenge to land a man on the moon when we did it in just 8 years and 2 months, a feat that seems ever more wondrous the more distant we grow from it, but it was even more than that. The Soviet Union had shown that success on the frontier of space could, and did, translate into power and influence in the world. In the cold war, we were in a strategic competition for just such power and influence against a totalitarian regime whose core values were abhorrent to most Americans.

So, when Americans watched the moon missions depart, our belief in freedom and in our way of life and our hopes for a better life for our children and their children were riding along with the astronauts. For a moment, we could leave our anxieties about the larger struggle of the cold war behind, and let our spirits soar into the skies. Nevertheless, we knew always in those years that we were locked into that struggle, that it was playing out most visibly on the space frontier and that we were finally winning. Because of this, the connection between what we do at NASA and the daily lives of Americans at that time was immediate and intense. Even more, these events were inspiring to the world, not just to the United States.

Friends of mine who have come here from other lands tell me directly that the world was cheering us on because of the sheer magnitude of the accomplishment that we attained. American self-confidence, our belief that we can do what we set out to do, drew admiration from around the world then as it does now. NASA, then as now, was the embodiment of that spirit.

Today, we are in a very different world. The military and political competition has largely receded into the background. Today, we are primarily engaged in an economic competition around the world. We increasingly live in a global economy where rising wealth and living standards also mean ever-heightened levels of competition from places we never even considered. There are now more software engineers in Bangalore, India, than in Silicon Valley. Japan, Taiwan and South Korea together generate a fourth as many patents as does the U.S. every year, and their percentage is growing rapidly.

The products of this innovation are all around in what has become a world marketplace. How many of you have a cell phone, a television or a car that was manufactured in the United States? These things are now a world commodity. I don't think I need to spend more time on these points. They are superbly treated in Tom Friedman's book, *The World is Flat*, and in the report, *Rising Above the Gathering Storm*, by my friend and colleague, Norm Augustine of the National Academy of Engineering. But I think the bottom line is that we all want our economy to continue to grow. We want to compete successfully. We want better lives for our children. But, economic growth and competitive success result primarily from the introduction of new goods and services or from finding more efficient ways to produce existing ones.

Economic growth is driven by technological innovation. Societies that foster it lead the pack, and others lag behind. But, if technological innovation drives competitiveness and growth, what drives innovation? There are many factors, but the exploration and exploitation of the space frontier is one of those. The money we spend, half a cent on the federal budget dollar, and the impact of what we do with it, doesn't happen out there. It happens here, and the result has been the space economy. So, if America is to remain a leader in the face of burgeoning global competition, we must continue to innovate, and we must continue to innovate in space.

There is another factor driving innovation, also, and in my opinion, it is too often overlooked, or if it is seen, it is too often dismissed. Success in an economic competition depends upon image as well as substance. Companies the world over have a choice as to where to do deals and with whom to do them. The nation that appears to be at the top of the technical pyramid has taken a very large step toward being there in fact. Developing countries like China recognize the value of space activities as the driver of innovation, a source of national pride, and a membership in the most exclusive of clubs, that of spacefaring societies. It is no coincidence that we are seeing thousands of high-tech jobs starting up in China.

NASA is uniquely positioned to drive the space economy with both substance and style because our mission requires us to push the technological envelope every single day, and to do it in the most publicly visible manner of any human enterprise. Our human and robotic ventures into the solar system, our attempts to fathom the mysteries of the universe require for their accomplishment a voyage of discovery beyond the limits of knowledge, and they are accomplished for all to see on a stage of breath-taking scope and grandeur.

At once, we have an endeavor which places the highest possible demands on technical ingenuity, requires a calculated, but stunning, audacity for its success and returns a product with which all of the world is fascinated. Even when we fail, and we do, we do so, in Teddy Roosevelt's words, "while daring greatly." That is why every year, the National Air and Space Museum is the world's most visited museum.

At NASA, we explore the frontier, and in fact, we create that frontier. To do it, we have to answer a lot of questions that wouldn't even have been questions without that commitment to the unknown. The answers to those questions are answers that power our future here on Earth. Because our mission is flight in all its forms in space and in the air, we think and work and do our engineering and science at the extremes, and that is where the discoveries are made.

In celebration of its own 25th anniversary, *USA Today* recently offered a list of the top 25 scientific breakthroughs that have occurred since its founding. Nine of those come from space, eight of them directly funded by NASA. We see the transformative effects of the space economy all around us through numerous technologies and life-saving capabilities. We see the space economy in lives saved when advanced breast cancer screening catches tumors in time for treatment using methodology developed from image extraction from the Hubble Space Telescope; or when a heart defibrillator restores the proper rhythm of a patient's heart; or when GPS, which was developed by the Air Force for military applications, helps guide a traveler to his or her destination. We see it when weather satellites warn us of coming hurricanes or when satellites provide information critical to understanding our environment and the effects of climate change. We see it when we us an ATM to pay for gas at the pump with an immediate electronic response via satellite. Technologies developed for exploring space are being used to increase crop yields and to search for good fisheries at sea.

All of this is very nice, but sometimes a personal example carries I think more weight than the most comprehensive set of factual data. So let's consider the case of a woman, Sarah Moody, and her young nephew, Steve, who was born with a very rare disorder. He had no sweat glands and he couldn't cool his body in the summer. He would overheat dangerously. After one too many close calls, Sarah thought to herself what many have thought and many have written: If we can put a man on the moon, why can't someone figure out a solution to this problem? So she called NASA and was put through to what is today our Innovative Partnerships Program. NASA scientists were able to adapt cooling technologies developed for the Apollo Lunar astronauts to develop a cooling vest for Steve. It worked. She started a foundation that has delivered 650-some vests to other people suffering similar disorders. Her foundation also turned to NASA for help with kids who had to live in dark rooms to avoid suffering tumors when exposed to ultraviolet light. NASA's contractors helped create suits that blocked the UV, allowing those kids to go outside. Sarah Moody died a few years ago, but her legacy lives on.

Gary Thompson was an athletic 50-year-old man with a family history of heart disease. He was given a clean bill of health in a series of tests with several doctors a few years ago and then had a heart attack while running in a marathon. He survived and subsequently heard of a new ultrasound imaging technology derived from algorithms used to process images of Mars at NASA's JPL. He was diagnosed correctly using the new technology, something all the other tests had failed to do. He was so impressed that he started a company to take the applications to market. Medical Technologies International, Incorporated, now makes this new technology more widely available. It is in use in all 50 states.

These examples only begin to tell the story. We can all be proud that they exist, but equally and in all fairness, we must recognize that we wouldn't create a space program in order to get these collateral benefits. But it is more than that. NASA and work in space generally are transformative. We don't just help develop technologies. We inspire whole new industries. We revolutionize existing ones, and we create whole new possibilities. In that vein, I often wonder if it might be possible to quantify the value to society of upgrading the standards of precision to which the entire industrial base of that society operates. Any company bidding on space projects—military, or civilian or commercial—any company who wants to be a subcontractor or a supplier, any company that even wants to supply nuts, bolts and screws to the space industry must work to a higher level of precision than human beings have ever had to before.

How do we value that asset? I don't know, but I am absolutely convinced that it is real, and that without the space industry, we wouldn't have it. In a related vein, another benefit to the space economy is the way that it inspires people to go into the technology sector. Our host today, Bob Stevens, was talking about exactly that experience of being a little boy watching a grainy black-and-white television set. Other people, Steve Jobs, Bill Gates and Burt Rutan tell similar stories. What is more important to realize is that a huge number of technology professionals in all fields first got hooked on space and then were inspired to pursue technical careers.

This is truly one of the best spinoffs we have, and the space exploration enterprise should receive due credit for it. At a time when we are concerned about declining enrollments in engineering and science and mathematics, this ought to be no small factor in our thinking. Space exploration inspires kids to study hard things, so that they can be part of it. Most of you know how the demands of spaceflight sparked the revolution in integrated circuitry. In the early years, our rockets couldn't compete with the throw weight of Russian rockets, and so in the United States, we embarked on a process to lighten the payload, and out of that, we got integrated circuits. But we didn't only get integrated circuits. We got all of the other technologies that make them possible. These capabilities now permeate our entire industrial base. The use of integrated circuits themselves is so ubiquitous in devices whose very existence would have been almost unimaginable only a few years ago that we no longer even notice it. Cell phones are given away as a competitive inducement to select one rate plan over another. Devices that can store gigabytes of information, a capability once beyond price, are given away as keychain fobs in promotional advertising. Built into your checkbook can be a calculator that Newton or Gaius would have given years of their career to have. For a few hundred dollars, you can buy a device that will allow you to navigate to any address in the country that can ever be found on a map. Who even notices anymore?

Today, NASA is again among those at the forefront of computational development, as evidenced by recent demonstrations of the first computer chips that can work at 500 degrees centigrade in very hostile environments. Or consider a recent demonstration of a quantum computer chip, a device that operates at the limit of our understanding of the physical universe and makes use of the very strange and elusive properties of quantum mechanics, properties that even physicist friends of mine themselves refer to as "quantum weirdness."

Quantum computing won't be just one more incremental improvement in present-day computing. It will revolutionize it. It is the kind of breakthrough you get when you set the bar possibly high, simply because the rigors of space exploration demand that it be so. To stimulate economic growth, to increase our international competitiveness, and to create better lives for our citizens, we must stimulate technological innovation. NASA's own programs accomplish this in one way, but as we have seen, the space economy today, at $180 billion around the world, is much bigger than NASA and is becoming more so. But NASA has another role to play, and that is as a catalyst for new ideas and new technology by setting extraordinary goals and then engaging the imagination and drive of entrepreneurs in the private sector.

One such effort is our program to enable the creation of new low-cost commercial space launch capability using as an anchor market the logistics requirements for the International Space Station. The COTS program, short for Commercial Orbital Transportation Services, is intended to demonstrate capabilities to provide low-cost transportation to orbit for cargo and crew. If this experimental effort is successful, NASA will purchase commercial services for delivery of cargo and crew to the station. We envision multiple flights per year beginning after shuttle retirement in 2010. To me, this is exactly analogous to the way that enlightened public policy spurred the aviation industry of the 20th century. That system that took us in 100 years from cloth, sticks and string to a transportation system where you are more likely to die from being struck by lightning than in an air transport accident. It is a stunning achievement, and we need to do it in space.

Fifty years into the space age, the greatest obstacle to the exploration and utilization of our solar system remains the very high cost of space transportation. No government effort has yet made a successful attack on the problem. But, when we do have it, we will find that commercially viable, low-cost space transportation will be as transformative to the economy as the transition from steam to diesel power or the achievement of powered flight that I spoke of a moment ago. It will open up possibilities that now appear impractical, if not outlandish.

And this takes us to the Vision for Space Exploration, laid out by the president in 2004 and codified in the NASA Authorization Act of 2005. In the wake of the *Columbia* tragedy, it calls for NASA to extend human and robotic presence to the moon, Mars and beyond. As the President's Science Advisor, Dr. Jack Marburger stated in his March 2006 speech at a Goddard symposium, "As I see it, questions about the vision boil down to whether we want to incorporate the solar system into our economic sphere or not." Precisely so, perfectly said.

Every aspect of human knowledge will be tested and advanced: physics, chemistry, biology and their practical applications in engineering, material science, medicine, computer science, robotics, artificial intelligence, power generation and storage and many other fields. I didn't even mention rocket science. This is a legacy that the crew of *Columbia* would be proud to know that we had carried forward. Reaching for the unknown, making our lives bigger and our horizons broader, achieving things never before possible are the heart and soul of what we do at NASA. By pushing beyond the future, by setting for ourselves seemingly impossible challenges, we are transforming our lives for the better here on Earth, even as we explore new worlds in space; and if, as Shakespeare said, life is but a stage, then NASA takes the play to the grandest possible stage. In doing so, we create the space economy. At NASA, we are making the future happen, and we are doing it now.

Thank you very much.

Partnership in Space Activities

Michael D. Griffin
Administrator
National Aeronautics and Space Administration

International Astronautical Congress
Valencia, Spain
October 3, 2006

I spent last week in Beijing and Shanghai, touring various facilities and meeting some excellent scientists and engineers. It was my first visit to China, and I will again take a moment to thank my hosts for their warm hospitality on that visit. It is important for the fraternity of spacefaring nations to discuss openly the issues that we each face. I look forward to more such dialogue with China's National Space Agency, and to continuing the dialogue with the heads of other agencies here at the International Astronautical Congress (IAC). Thus, my remarks today will be on the subject of "partnership" as we apply it to our endeavors in space.

Space exploration, whether human or robotic, is still the grandest and most technically challenging expression of human imagination of which I can conceive. Thus, I believe it to be in our best interests in this unique human endeavor to work together on occasion, to ask each other as different countries and different cultures how we should go about solving the unique problems of this unique endeavor. The physics is the same for us all; the rocket equation does not change when expressed in a textbook of a different language. But I have found that while the problems and the physical constraints are the same for all, the vagaries of human creativity and ingenuity can yield many different solutions. So, it really is to our mutual benefit to understand how each of us develops the art and science of spaceflight. We all have much to learn. We can learn best by doing some things together.

I have often said, but it bears repeating before this audience: I have no doubt that humans will continue to explore space, going to the moon and Mars and far beyond. Thus, the question of "whether" this will happen is not an interesting one to me; I know that it will. The interesting questions center around topics like "when," and "who," and "what," and "why." When will humans next return to the moon, or venture to Mars, or first explore the near-Earth asteroids? Who will first do each of these things, and many even bolder things beyond them? What languages will they speak, and what values will they hold? Why will they go; what gains will they expect to return to their parent societies?

Such questions can be considered jingoistic if taken out of context, but that is not my intent at all. My intent in raising them is to ask how each of our cultures regards its role in exploring the space frontier. The American culture retains even now a certain frontier mindset, based on our history. We in America are the descendants of pioneers from Spain, Portugal, Holland, Great Britain, France, Germany, and many, many other countries who emigrated over many generations to settle in what became the United States. But the British were the boldest and most persistent of these early groups, and so the primary language of the United States came to be English. Canadians speak both English and French, while elsewhere in the Americas both Spanish and Portuguese are spoken. Now, these various languages not only convey the thoughts of their speakers in different ways, they also encourage and allow different thoughts. Language is, in part, a window into, and a map of, the culture of its users. And so, looking into the future of space exploration, I sometimes wonder what languages the explorers and eventual settlers of the moon and Mars will speak? Will my language be passed down over the generations to future lunar colonies? Or will another, bolder or more persistent culture surpass our efforts and put its own stamp on the predominant lunar society of the far future?

Further, the laws of the United States, which represent the values of our people, are fundamentally based on English common law, Roman law and the Justinian code, yet have evolved to take into account modern philosophies and practicalities. Especially noteworthy is our core belief in the possession by individuals of certain inalienable rights, including the right to own property. All of the countries represented here at the IAC are governed by the rule of law, but each of us has variations in our legal codes which reflect the values of our unique cultures. So, looking into the future of space exploration and space settlement, what values and laws will govern those explorers and settlers?

These and others are fundamental questions which I will not attempt to answer here, because in the end I am not qualified to do so. I have never pretended to be either a linguist or a lawyer; I am merely an engineer. However, I consider such topics to be quite fascinating, and I hope that the community of spacefaring nations will carefully consider their import in the future. While we may disagree on certain points and priorities, it is important that we try to understand, and respect, each other's views. This is an essential ingredient of any successful partnership.

It is no secret, and should be no surprise, that the United States has played, and seeks to play, a leadership role among the community of spacefaring nations. But we cannot simply presume such a role; it can only be earned. We must first be respected as a good partner before we can be regarded by you, the community of spacefaring nations, as a good leader. We at NASA have not always been the most reliable of international partners, and it has been one of my most important goals to improve that record. All who are here know that I have said on many occasions that the partnership behind the International Space Station provides its highest and longest lasting value, a value which we in the United States highly respect, as we work with our partners toward the completion of this enterprise.

For many and various reasons, partnerships in space exploration have enormous benefits, but they are not easy to consummate. History demonstrates that countries and cultures will always have issues which divide and set them apart. We compete in the global marketplace of ideas, influence and intellectual property as well as in the more visible marketplace of economic goods and services. It should be no surprise that there are sometimes disputes surrounding one or more of these issues.

Competitiveness is healthy and useful for people, organizations and even nations, but the competitive spirit must be leavened with a healthy dose of collaboration, lest it be carried past the point of utility and into harm. So, while competing, we need also to be mindful of opportunities to work together, to create alliances for the common good of mankind. I believe that space exploration and scientific discovery are examples of endeavors which offer a distinctly unifying force for that common good.

However, each of our countries also has unique national security concerns. Having spent a good portion of my career working for the U.S. Department of Defense, I am not ignorant of the military applications of space technologies, nor of the need to regulate the proliferation of certain capabilities, and missile technologies are prominent among these. The United States is firmly committed to ensuring that certain key technologies, which we possess and some others do not, not be used against us or our allies. That priority is higher for us than partnership in various space endeavors, and this fact must be understood and carefully considered by the parties involved in any putative collaboration. I recognize the bluntness of this assertion, but I believe that each of us, as spacefaring nations, must respect each other's national priorities, and must speak openly and honestly with each other if there are differences which hamper our ability to collaborate.

Further, each of our countries has only so much money to expend on space endeavors, and this also limits our ability to partner on various projects. Even with an annual budget of $16.8 billion, NASA cannot afford everything that our own numerous constituencies would like us to do in exploration, science, and aeronautics research. That budget constitutes only 0.6 percent of the overall budget for our U.S. government, but in the wake of Hurricane Katrina, the greatest natural disaster in the history of the United States, and the expense of the Global War on Terrorism, I still consider myself to be very lucky that our nation's leaders provide that much to NASA. But still, we must carefully choose those endeavors to which we commit with our fellow spacefaring nations. Much as we would wish otherwise, we cannot do everything we would like to do. In this context of limited resources, it is clear that partnerships work best when all partners have "skin in the game," each contributing resources toward a common goal that is greater than that which could be easily afforded by any single partner. We believe that such relationships work best when conducted on a "no exchange of funds" basis. I must admit that this view is not universally shared. On many occasions since assuming my role as administrator I have been asked about opportunities for "partnership," when what is really being sought is American investment in the aerospace industries of other nations. I must be clear on this; "partnership" for us is not a synonym for "helping NASA to spend its money."

The United States' Vision for Space Exploration honors our past commitments to the space station partnership, and calls on your interest and support in embarking upon new ventures. Last month, we restarted assembly of the station, after a hiatus of over 3 years due to the loss of Space Shuttle *Columbia*. On board the space station today, American Michael Lopez-Alegria (who was born in Spain but grew up in California), Russian Mikhail Tyurin and German Thomas Reiter are part of the greatest construction project in the history of humankind, rivaling the pyramids of Egypt, the Suez and Panama Canals, or the Great Wall of China. Who would have imagined after World War II, my own father's generation, that such a team could be working and living in space today?

Two weeks ago, I welcomed home the crew of Space Shuttle *Atlantis*, which included Canadian astronaut Steve MacLean. Last week, Russian cosmonaut Pavel Vinogradov, American astronaut Jeff Williams and Iranian-born spaceflight participant Anousheh Ansari returned home to Earth on *Soyuz*, landing in Kazakhstan. In December, I look forward to the next shuttle assembly mission to the space station, with Swedish astronaut Christer Fuglesang. Even at a cursory look at human spaceflight activities over just this past month shows that space exploration is a truly international endeavor, and a broader look shows just how true this has been, for a long time.

The shuttle program has, in 25 years of operation through STS-115, flown 708 astronaut-seats. (By that I mean that most individuals have flown more than once.) Eighty-three of these flight opportunities, or about 12 percent of the total, have gone to 58 individual international astronauts from 14 countries. I don't know the statistics for *Soyuz*, but I do know that our Russian partners have flown a substantial number of non-Russian cosmonauts, going back for decades. This goes beyond the mere exchange of money or favors or other considerations. The largest spacefaring nations have quite simply made it a point to make human spaceflight a significant international activity.

But that's not all. Last month, American engineers and scientists met in India to review progress in executing data-sharing agreements and delivering two instruments for India's Chandrayaan lunar mission. Since NASA cannot afford to do everything, and since so many missions are planned for the moon over the next few years, including China's Chang'E mission, lunar science data should be openly shared among the science community, just as we do with other planetary science data.

Also last month, the Japanese Space Agency successfully launched the SOLAR-B satellite, a joint JAXA-NASA-UK-Europe heliophysics mission to study the sun's magnetic fields. Not only is NASA interested in the sun's effects on terrestrial telecommunications and power grids and potential impacts to the space station, we'll soon need timely and accurate warnings of impending solar storms for our astronauts in cislunar space. Later on, we'll need this same information on treks to Mars and near-Earth asteroids.

Later this month, I'll be sitting down with the NASA management team to go over the flight data from the past three shuttle missions to see if a servicing mission to the Hubble Space Telescope to extend the life and capabilities of this Great Observatory can be performed safely. People from all over the world are awed and inspired by Hubble pictures that reveal the secrets of our universe. From the first, Hubble has had an international complexion. After that mission is completed, the European Space Agency will launch the U.S.-built James Webb Space Telescope aboard an Ariane V. We're also collaborating with the German Aerospace Center on the SOFIA airborne observatory, and along with NASA's Kepler mission, ESA's Gaia mission will survey our galaxy for extra-solar planets.

Next month, NASA Deputy Administrator Shana Dale will meet with her counterparts in Washington at an AAS/AIAA seminar to define better our purposes in going to the moon, and to discuss what we'll do when we get there. This meeting is a follow-up to a highly successful NASA-hosted workshop last April. Since then, we have met regularly with other international space agencies to define a global strategy for space exploration.

One aspect of this discussion is the need to set certain engineering technical standards to ensure compatibility and interoperability in our exploration architecture. Analogous to my previous comments about spoken languages for future space explorers, it is important that the engineering standards for NASA's architecture be specified with the international metric, or SI, standard as the base unit of measure, with English units only by exception when it makes sense for NASA to do so. Thus, we hope for a high degree of compatibility of interfaces and standards, as spacefaring nations explore the moon, Mars and near-Earth asteroids together.

So, before I open up the dialogue to your questions, let me share with you the awe that veteran American astronaut Shannon Lucid conveyed to me last week as we toured China. Her parents were American missionaries in the city of

Shanghai, and Shannon was born there during World War II. Her family was interned in a concentration camp for the first year of her life, after which she and her parents were released as part of a personnel exchange. They returned to China after World War II, and she attended kindergarten there. She has many memories from that time. She was amazed last week by the transformation of the city of Shanghai from what she remembered from the 1940s. Such changes are never objectively surprising, yet when we are confronted with them, as individuals we are indeed always surprised.

Cities change; people change; nations change. Some nations that were American allies during World War II are not as close to us today, and some nations that were enemies in that era are now among our closest partners. Many have asked why I visited China last week on behalf of my country, when that nation is today not among those most closely linked to us. But China is a powerful and important nation, home to the oldest civilization we have in the world. The United States is newer and younger, but it is also a powerful and great nation. There is no possible purpose to be served by creating or advocating adversarial relations between the United States and China, or indeed between ourselves and any other nation. There have been sea changes in relationships between the United States, Germany and Japan, our adversaries in World War II, and between the United States and Russia, our competitors during the Apollo era of the 1960s. There can be more such changes, and there will. The best possible goals for those who manage our nations' space agencies are to find ways to narrow the differences between us, so that the changes are good ones. We need to look toward those things we have in common, precisely because there are already an ample number of things to divide us. Perhaps this is not "breaking news" so much as it is a new perspective on the news.

Thank you.

International Space Cooperation

Michael D. Griffin
Administrator
U.S. National Aeronautics and Space Administration

Parliamentary Group on Space
French National Assembly
June 5, 2008

President Accoyer, distinguished members of the Parliamentary Group on Space and guests, thank you for the opportunity to address you today. I am honored to be with you and look forward to the opportunity to discuss bilateral cooperation with France as well as more general European space collaboration with NASA, and my perspective on the prospects for the future of space exploration. I firmly believe it is a future that will be full of opportunities for meaningful collaboration, but that achieving it will take hard work and determination. With France assuming the Presidency of the European Union this year, and with a European Space Agency Ministerial in November to help define the European Space Agency's (ESA's) future direction, I am aware that I am speaking to you at a crucial time, when France is considering how best to exercise its leadership in space to benefit its citizens and the broader European community.

As has been the case with the Centre National d'Edutes Spatiales (CNES), we have enjoyed excellent relations and have had the privilege of robust civil space cooperation with ESA, the German Aerospace Center (DLR) and the Italian Space Agency (ASI). As we proceed with space exploration we look forward to continuing to work with France, Germany, Italy, the European Space Agency and, indeed, with all of our European counterparts.

Allow me to begin by highlighting a very important aspect of relations between France and the United States, which is simply that France has long been the United States' closest partner in space exploration, by many measures. For example, NASA has more active space cooperation agreements with France than any other country. This is a relationship that goes back to the earliest days of spaceflight. In fact, I could note that NASA's first overseas representative was deployed to Paris in 1964. That is a position that we have filled continuously ever since, and our base of operations in Europe remains in France. In January 2007, then-Minister for Higher Education and Research, Francois Goulard, and I signed a U.S.-France Umbrella Agreement for Cooperative Activities in the Exploration and Use of Outer Space for Peaceful Purposes. In signing this agreement as the foundation for further space cooperation between our two nations, Minister Goulard and I underscored the vital role that a robust program in space exploration can play in sustaining our respective economic and national security interests, promoting innovation, and motivating pursuit of and excellence in mathematics, science, engineering and technology. I would like to thank the leadership of this Group, and of the National Assembly, for ensuring that this agreement was ratified.

NASA's cooperation with France touches almost every aspect of our activities in aerospace.

In aeronautics research, NASA is working closely with the French National Aerospace Research Center and others to develop a better understanding of the issues associated with aircraft in-flight icing. These research efforts will ultimately improve ice accretion modeling techniques and refine ice detection instrumentation and measurement systems.

In space science, in addition to a long history of successful cooperative planetary missions including a robust program of Mars exploration, a variety of French institutions and scientists are important contributors to NASA's Gamma Ray Large Area Space Telescope (GLAST) mission. The GLAST will, among other things, help us to study black holes and the source of gamma-ray bursts—the most powerful explosions in the universe—to probe dark matter and the early universe and to explore early star formation, pulsars, solar flares and the origins of cosmic rays. Cooperation in future space science missions is also being considered in the study of dark energy. Closer to home, NASA is looking forward to the upcoming four-party Ocean Surface Topography Mission (OSTM), scheduled to launch next week. This mission, involving the National Oceanic and Atmospheric Administration (NOAA), NASA, CNES and Eumetsat will continue a long line of cooperative U.S.-French Earth science missions, beginning with TOPEX/Poseidon and continuing with Jason-1 and CALIPSO. These successful cooperative missions have provided us with an improved understanding of ocean circulation; weather; climate variability and air quality; and they clearly demonstrated an effective model for transitioning research and development activities to practical applications on a day-to-day basis.

These few examples illustrate the significant benefits of our cooperative relationship, one that has produced benefits not only for France and the United States, but also for millions of people around the world. I have no doubt that such fruitful scientific collaboration will continue into the future.

Another subject that may continue to be of particular interest to you is cooperation in human spaceflight. In this regard, I must note that NASA and the International Space Station partnership have come a long way since the tragic loss of Space Shuttle *Columbia* in February 2003. That accident forced the U.S. government to reconsider the strategic reasons for human spaceflight. As a result, the United States has committed to a long-term program of human and robotic exploration of space for a variety of purposes, purposes that I believe we share with Europe.

When I look at the European Space Policy document released last May by ESA and the European Commission, and when I read the report by Senator Revol and Dr. Cabal published last February, or I review the remarks made by French President Nicolas Sarkozy on space policy given earlier this year in Kourou, I see that we have much in common. We both see space as a strategic environment that generates multiple advantages for our economy and our people. We see it as a means of promoting peaceful international cooperation. We see that space exploration, in general, and human spaceflight, in particular, energize and encourage our minds as does no other enterprise. We see that it inspires our children to study math, science and engineering so that they can be a part of this great endeavor. As President Sarkozy noted, access to space is the hallmark of major industrial and technological powers.

For these reasons, which you understand as well as anyone, the United States will never turn its back on this great endeavor. So, in the aftermath of the tragic *Columbia* accident, the United States committed to a new space policy. In my opinion, it is the best space policy we have ever enunciated. It builds upon our successes, pulls the components of our space efforts together into a more integrated whole and keeps faith with our long-standing partners like France. In the field of human spaceflight, we decided that it was time to replace the 1970s-vintage space shuttle with a new vehicle that will be capable of taking us beyond low Earth orbit, be safer and cheaper to operate and be flexible enough for our children and grandchildren to use when they head for Mars and other destinations. The Orion crew vehicle and Ares launch vehicles are being developed today with these goals for the future in mind.

The first phase of this new exploration policy is to complete and operate the space station together with our international partners. I am thrilled with the excellent progress being made toward that objective. Beginning with the return to flight of Space Shuttle *Discovery* 3 years ago, the shuttle has conducted nine successful missions to the space station. While they have continued to become, technically and logistically, ever more complex, they have been executed brilliantly, thanks to the skill, dedication and hard work of the people involved.

ESA Astronaut Léopold Eyharts, is one of those people. A credit to the ESA Astronaut Corps and the French Air Force, General Eyharts flew to the station on Space Shuttle *Atlantis* earlier this year, became part of the crew of Expedition 16 and returned to Earth on Space Shuttle *Endeavor* in

March. While on the station, he tested and operated the European *Columbus* Module, marking a significant milestone both for Europe and the partnership. The culmination of many years of effort in Europe, *Columbus* gives Europe the capability to conduct onboard research in areas such as material science, fluid physics, life science and Earth observation. It is a major contribution to the station and demonstrates the maturity and sophistication of the European space enterprise.

With respect to the recent launch of the Automated Transfer Vehicle (ATV) cargo vehicle earlier this year, let me re-emphasize what I said on April 3. Like many of my colleagues, I was happy to see that the first ATV was named after the French writer who was such an inspiration to so many of us in the space business. I am incredibly proud of our European partners for successfully docking the Jules Verne with the station. I applaud Europe's achievement. Only the United States and Russia have previously conducted automated dockings in space. In combination with the launch of the *Columbus* Module earlier this year, the success of the ATV marks the arrival of Europe as a full-fledged space power. Now that the ATV is operational, we hope Europe will utilize it to its fullest potential, providing NASA and the other international partners with cargo capabilities through existing arrangements. Commercial services, like the ones I am working to foster through NASA's Commercial Orbital Transportation Services (COTS) program, could follow.

We await with anticipation the many deliveries by the ATV of critically needed cargo and equipment to the station. Further, the many technologies developed throughout Europe for this sophisticated vehicle offer the prospect of even greater European feats in the future, based on the use of this core vehicle. It would be a small step from today's Ariane 5 and Jules Verne to an independent European human spaceflight capability. In the meantime, the ATV is a tremendous asset for Europe in space, and we expect you will make the most of its capabilities for years to come.

The space station is on schedule to double its crew next year and to be completed in 2010. For those who have heard that the U.S. will soon abandon the station, let me be clear: we are committed to building and utilizing the space station well into the next decade. In fact, in the new U.S. space policy, the station is the primary focus of our near-term effort. Human research on the space station will directly benefit our understanding of and preparation for future activities on the moon and later voyages to Mars. Further, the U.S. Congress has designated the U.S. segment of the station as a national laboratory and directed NASA to develop a plan to "increase the utilization of the station by other federal entities and the private sector. … " Congress does not create or eliminate national laboratories lightly. Thus, it is inconceivable to me that the U.S. would abandon a perfectly functional space station because we have arrived an arbitrary date on the calendar. So while I cannot speak for a future U.S. administration or Congress, I do believe that the station will be around for a long time beyond 2016, and the U.S. will remain part of it.

As I noted earlier, present U.S. space exploration policy was born of the *Columbia* tragedy. One of the findings of the *Columbia* Accident Investigation Board was that "the loss of *Columbia* and her crew represents a turning point, calling for a renewed commitment regarding exploration." In the U.S., the plan for space exploration was put forth by President Bush in January 2004 and, after nearly 2 years of informed debate, was ratified with a remarkable level of bipartisan support with the passage of the NASA Authorization Act of 2005. Republicans and Democrats in the United States may disagree on many things, but they found common cause in the development of a coherent space policy.

After embracing our international commitments to complete the station, the policy directs us to extend human and robotic presence throughout the solar system, first to the moon, and then on to other destinations, such as Mars. At this point, I should emphasize that as part of this remarkably straightforward policy, we are also directed to seek opportunities for international collaboration in our activities. I will discuss this point in more detail later.

But before I do that, let me take a moment to address a few of the concerns I've heard about the path the U.S. has chosen with regard to exploration. Some in France at the time of the president's announcement asked what France could gain from an American desire to return men to the moon. More recently, some in Europe have suggested that any consideration of missions to the moon is a distraction from the real goal of a human flight to Mars. Others have predicted that significant change to U.S. space policy will come with the presidential elections, so it is best to sit on the sidelines and wait to see what the next U.S. administration will do. I have several observations to make in connection with these points.

First, U.S. civil space policy is specifically designed for the long term, designed to be implemented affordably and systematically across many changes of administration and Congress. In fact, those who are in favor of continuing human spaceflight, a substantial majority of U.S. policymakers, agree that we have little choice but to proceed on the path we are now following. While there will certainly be debate on the details of NASA's plans, in my view there will not be a significant change in our overall direction. There is a broad bipartisan consensus of support for today's U.S. civil space policy.

Second, after nearly 30 years of service, the space shuttle will cease operations in 2010. The shuttle is an aging, fragile and increasingly expensive vehicle to operate. Production lines are in the process of shutting down. Suppliers are no longer are making certain critical parts. Vendors are moving on to other businesses. Meanwhile, our Orion spacecraft and Ares I launch vehicle will soon come on line. The major Orion and Ares I components were put under contract last year. An engineering model of the Orion vehicle has been built and will be used to test the launch escape system this September. The first Ares I test flight, Ares I-X, is scheduled for mid-2009.

Finally, I should note that although the "gap" between the shuttle and Orion will be painfully long, the shuttle will be replaced by vehicles with significantly higher reliability and flexibility. By the middle of the next decade, multinational crews will be traveling to and from the station—six at a time—in the Orion spacecraft. We are committed to the new path.

That still leaves us the question of destinations for human exploration. I wholeheartedly agree with President Sarkozy that Mars offers us a great adventure, and I also agree with those who say that Mars is the ultimate destination for mankind in the 21st century. But I do not believe that Mars is the only interesting destination for mankind in the inner solar system, nor do I believe that it is reasonably within our immediate reach. As Professor Stephen Hawking said in his April lecture at the George Washington University, Earth's moon is an obvious first stop in human exploration, because of its proximity and the potential that water ice may exist in its polar craters. He also noted that Mars is the obvious next target after the moon. I agree with him completely.

In the United States we had similar types of discussions in the early 1960's, when we first engaged in the race to the moon with the Soviet Union. At that time there were numerous debates about the surest path to our destination, debates which could not be resolved with only the experience accumulated during Project Mercury. We had capabilities to demonstrate and technical skills to hone before we could go deeper into space. The Gemini Program helped us to develop the needed skills for the Apollo lunar missions, while remaining safely in our own "neighborhood," which at that time was low Earth orbit.

Similarly, as we look toward Mars, continued work in Earth orbit followed by "field tests" on the moon are, in much the same way, on the path to a successful human expedition to Mars in the next few decades. During the development of U.S. space exploration policy, consideration was given to going directly to Mars. But when we looked at it carefully, we decided that we could neither afford nor sustain the budget increases needed to ameliorate the risk of such a plan.

Let me turn now to the subject of international cooperation in exploration. It is sometimes said that NASA was overly prescriptive of the roles of its partners in the early days of space station development. I think much might be said on either side of this claim and has. But whether or not it was so in the past, I determined that it would not be the case in the future. So, early in my tenure at NASA, I stated that we will not attempt to prescribe the manner of participation of any of our potential partners. We will work with others to define an exploration architecture suitable to all, and we will identify those portions of the task that we are willing and able to accomplish with the funding we can provide. We expect that others will do the same.

In the years since, the exploration architecture is coming together nicely, with broad international support. NASA has welcomed ideas from our friends in France, elsewhere in Europe and from many other countries. We will continue to do so as we move forward. Just as with mid-20th century Antarctic research, some of the most creative approaches to 21st century lunar and Martian exploration will depend on international collaboration. France, through its space agency CNES, has added its voice to those of 13 other space agencies in a multilateral dialogue we collectively refer to as the Global Exploration Strategy. This effort has gone so far as to publish a framework document and establish a coordination mechanism, called the International Space Exploration Coordination Group. Exploring the moon, and eventually Mars, will be a challenging task, one that NASA has neither the resources nor the desire to do alone. And, as I mentioned earlier, U.S. policy and law explicitly calls on NASA to engage in international cooperation in pursuit of our goals. As European experts and political leaders have noted, a global exploration effort is a key to unlocking the door to our future.

One other curious comment that I have heard in international discussion about U.S. space policy is the suggestion that the United States is somehow unfairly excluding international partners from the development of Orion and Ares. I say "curious" because I note that Senator Revol and Deputy Cabal in their report last year stated that "autonomous and competitive launchers" are an absolute priority. President Sarkozy said earlier this year that independent access to space was essential. The "vital importance" of "independent, reliable … access to space …" is also enunciated in last May's European Space Policy statement. I fully agree, and moreover, such independent access to space is of no less importance to the United States. That is why we are proceeding with these developments as national projects, while at the same time hoping that other aspects of exploration will offer fruitful soil for international collaboration.

Let me be clear: while we know that our national capabilities allow us to reach the moon again alone, we would not consider that to be a successful outcome. Measured against the standard of our own policy, it would be a failure. A group of nations pursuing common, coordinated goals will achieve so much more than a single country's mission or outpost.

I am personally committed to the idea that this enterprise should be international in scope. It is obvious to me that we share a commitment to international cooperation of this sort. For example, ESA's Space Exploration Policy Advisory Group noted as far back as 2004 that "the cooperation objective among key actors should be based on heteronomy, partnership and networking." In our lunar ambitions, we couldn't agree more. We prefer a coordinated effort at the moon involving many national space programs over the alternative of exploring space alone. We are trying to behave in a manner which supports that claim.

So NASA, guided by the U.S. policy, is pursuing a path of international cooperation in its space endeavors. It is a path that differs significantly from the Apollo era, and builds on the successes of prior shuttle missions and the space station. We welcome, indeed we are asking for, European collaboration in human exploration. We welcome the development of independent European capabilities in space to provide redundant systems in the event of failure of any one partner's capabilities. Between and among us, we have seen enough such failures that we should know by now to plan for them. We think that this would be of benefit both collectively and individually, especially if we can link individual capabilities via common interfaces that will ultimately provide the robustness we need for future ventures beyond low Earth orbit.

The foundation of NASA's cooperation with all of its partners is based on the principles of transparency, reciprocity and mutual benefit. Our relationships with our long-established partners in Europe and around the world have shown that only this foundation can provide a reliable basis for cooperation in spaceflight, and this foundation would be a necessary precondition for any new relationship with countries such as China.

In November of this year, the ESA Ministerial will make programmatic decisions regarding Europe's plans in space for the next 3 years and impacting ESA's direction for a much longer time by setting the stage for the 2011 Ministerial. This period, between 2008 and 2011, will be an important time for all of us. NASA will be working to enable early lunar exploration, following a stepping stone approach on the way to Mars and beyond. In Europe, these next years will be important for defining European objectives and putting in place the activities necessary to meet them. I hope that the decisions made at the 2008 Ministerial will hearken back to those of 1985, when the ESA Council agreed to pursue cooperation on the space station program.

Regarding past decisions, we are pleased that European nations came together in 2001 and displayed their commitment to long-term robotic and human exploration with the initiation of the Aurora Program, which targets potential human presence on Mars in roughly the same timeframe as U.S. space exploration policy. We welcomed the generation of momentum in Europe toward human and robotic space exploration in the European Commission's plan for implementing European Space Policy in 2003, and ESA's publication last year of objectives and interests in space exploration. In the future, I hope that you will maintain this momentum by encouraging the ESA Ministerial of 2008 to commit to a program of synergy and common purpose, that will bring our programs closer together, and that will allow us to leverage our limited funds to mutual benefit.

In summary, we thoroughly enjoy our productive relationship with France in space activities, both through bilateral and multilateral agreements with CNES and through French membership in ESA. We understand that there are many reasons to invest in the noble goal of space exploration, but that the same reasons will not have the same weight for all participants, and that other nations will embrace similar goals from different perspectives. In 2004, Jacques Blamont wrote that in the world outside the U.S., the decision of any government to spend money for space programs is motivated by societal factors, and not only to "fulfill the public's sense of destiny" as some have said about the U.S. space program. I agree with Professor Blamont on this point. France will identify its own rationale for pursuing space exploration—one that meets its national goals and the needs of its unique society, just as we do. France will carry those ideas forward to its partners in Europe and beyond. But in the end, I believe that we share common purposes and goals, and that it makes sense to pursue these mutual interests together, as France and the U.S. have done in space for the last 50 years. I sincerely hope that this will be the case for decades to come.

Not long ago, those of us in this business questioned whether we would ever again leave low Earth orbit. But now the question is not whether humans will extend their presence throughout the solar system. The questions are who will do it, how and when will it be done? I believe we have a firm handle on defining the "how" and the "when." As an engineer, I understand these variables. What I cannot control is the "who." Europe's decision on whether or not to pursue an ambitious program in space exploration is, of course, a decision that only Europe can make. Europe certainly has the capability. I believe it has the ambition. And I believe the French people have the visionary leadership to influence the rest of Europe to choose a path of partnership with the United States that will benefit us all.

I hope my words today have given you cause to consider this question and to join us and other spacefaring nations on what I believe will be the greatest of human adventures.

Thank you.

National Strategy and the Civil Space Program

Michael D. Griffin
Administrator
National Aeronautics and Space Administration

National Space Symposium
U.S. Space Foundation
Colorado Springs, Colorado
April 12, 2007

Today I want to discuss in a bit more detail a theme to which I have alluded in many talks, and that is the strategic importance of the civil space program to our nation. This is not a topic that receives a lot of attention. It is considered obvious to all that the space activities of our military forces and the intelligence community are "strategic." We talk about "strategic missiles" and "strategic reconnaissance," and the Russians make no bones about it with their "Strategic Rocket Forces." But civil space? Isn't that simply about scientific discovery, human exploration, or practical applications such as weather monitoring, navigation and communications?

I think it's not "simple" at all, actually, so let's talk about it.

Prominently featured in the Denver International Airport is a statue of a space-suited Apollo astronaut, the late Jack Swigert, a Denver native. A similar statue occupies a place of honor in the U.S. Capitol, one of the two allocated to each state to honor great men or women who represent the history and ideals of their home state. When people tour the Capitol, this is a statue at which they stop for a moment. Other exhibits represent our nation's past, and the present is captured by the view of our nation's current leaders hustling past, on their way to cast votes or attend hearings. But when tourists see this statue, they are arrested by the realization that they are glimpsing the future, not only that of our nation, but of the human species.

Jack Swigert grew up in Colorado, earned a mechanical engineering degree from Colorado University-Boulder, joined the Air Force and flew combat in Korea. He left the Air Force to earn a master's degree in aerospace engineering and became a test pilot for North American Aviation. He was selected for the astronaut corps in 1966, in the group that, somewhat tongue-in-cheek, Apollo 11 astronaut Michael Collins dubbed the "Original Nineteen." This group was selected at a time when it was thought that we would be conducting many more Apollo missions than regrettably turned out to be the case; several had to wait to fly for the first time on the space shuttle.

In April 1970, Jack replaced command module prime pilot T. K. Mattingly when the latter was exposed to German measles and flew his only space mission, Apollo 13, with Jim Lovell and Fred Haise. Both the flight crew and the ground controllers demonstrated their bravery, perseverance, and quick thinking again and again as they struggled to survive and return to Earth. This is the kind of thing that has caused me to say that those of us in the space business must live by a creed of excellence, or die from the lack of it.

The Apollo 13 mission was dramatized in a movie a few years ago. Jack Swigert was portrayed by actor Kevin Bacon. Unfortunately, in our culture today far more people recognize the actor than the man he portrayed, and far more people will flock to a movie depicting dramatized bravery than can recognize the real thing.

Anyway, a few years after the Apollo 13 mission, Jack left NASA to become the staff director for the House Science & Technology Committee and then returned to Colorado and was elected to Congress in 1982. He died of cancer before he could take office, at 51 years of age.

Now, Jack was not known as a perfect person; his time on the national stage was brief; and that little was suffused with professional disappointment and personal tragedy. So why did the people of Colorado choose him to represent them in the Capitol? Today, Colorado remains one of the most beautiful of the states, which, a century and more ago, were on the western frontier. In many areas it is unchanged from that time, a land that in places can still be seen as the mountain men saw it. The American West is no longer a frontier, but the people who live here can still see it from where they stand. Certainly the people who sent U.S. Army Captain Zebulon Pike west to Colorado, where he discovered the peak to the west of town that bears his name, fully understood that the exploration and development of the western frontier was a strategic issue. And I believe that Westerners today, more than most, understand viscerally that our nation's next frontier lies 200 miles above our heads.

So I think that the choice of a statue of Jack Swigert to represent Colorado in the halls of the U.S. Capitol was perfect. I believe it was because those making the choice understood the real reasons why we're in the space business. They understood that, for America, exploration is a matter of national strategy.

So, out of respect for the people who recognized the strategic importance of opening the American frontier, or those who built some of the great feats of engineering we enjoy today, let us ask ourselves some fundamental, and disconcerting, questions: Do we really understand the importance of what it is that we choose to do, or not, in space? If our great-great grandparents accepted the challenge of expanding the frontier of their time, will our generation do less? And if so, why?

NASA is a nearly unique government agency in the sense that it enjoys enormous name recognition and immensely positive public approval, consistently 65 to 75 percent as measured in professional surveys. This is a level of popularity that any public figure would envy, a level of "brand loyalty" about which most commercial product marketers could only dream. However, only about 50 percent of the people surveyed believe NASA to be relevant to their lives. So, in effect, the same people who resoundingly approve of NASA are not sure why. But when those being questioned are informed of even some of the more prosaic contributions of the space program to their daily lives—things like the development of integrated circuits, medical monitoring equipment for hospital patients, navigation and weather satellites, materials used in joint replacement surgery—their assessment of our relevance shoots above 90 percent. Collaterally, the approval rating for space exploration jumps from 70 to 80 percent.

So, clearly, the American people broadly approve of NASA even while admitting that they do not understand the relevance of the space program to their lives, and their approval increases further when we give them concrete reasons for it. To me, this is an extraordinary result. How can it be?

I have begun to believe that NASA's and the space program's place in the American consciousness lies not in our minds but in our hearts. The space program embodies in many ways what it means to be an American, the things we care about once we've dealt with the basics of earning a living and providing for our families. NASA's endeavors invoke feelings of national pride, what remains of American idealism and hope and innovation and daring, and respect for those qualities. And, yes, when they don't turn out well because we are human and therefore flawed, our endeavors also remind us of the need for determination, courage, resilience, toughness, and persistence and of respect for those qualities as well. Feeling for NASA involves a sense of our place in the world, of the need to pass on a legacy for our children and grandchildren, the hope that they will live in a better world, or maybe even on new worlds. Feeling for NASA involves the deep satisfaction of overcoming the most demanding technical challenges known to man. And, yes, feeling for NASA invokes the concrete benefits we obtain for our entire society when we tackle, and learn to overcome, those challenges.

Tom Hanks, who starred in the movie *Apollo 13*, and told the story of Apollo in the TV mini-series *From the Earth to the Moon*, speaks eloquently of what NASA's missions to the moon meant for him and our nation during the tumultuous 1960s and early 1970s with the Vietnam War, the civil rights movement and the assassinations of John and Robert Kennedy and Martin Luther King, Jr. In a simple, yet fundamentally insightful way, Hanks said of the Apollo program: "If we can do this, we can do anything."

I believe this thought provides more of a justification for our space program than any rational, dollars-and-cents explanation I can ever hope to provide as to what NASA represents to the American public and those of us in the space business. The Apollo program became the standard by which future feats of engineering and the focus of national willpower would be measured. "If we can do this, we can do anything."

However, a dark cloud passes over this bright thought. It has been a long time since we did "this." It has been over 35 years since man last set foot on the moon. Several of those who made that journey are no longer with us, and more will have passed before we return. While reading a recent story in *The New York Times* on the impending retirement of the space shuttle and its effect upon long-time space watchers in Florida, for whom the shuttle has become a fixture of daily life, the reporter noted that some young people today actually question whether we ever really achieved the goal of which President Kennedy spoke so eloquently: "landing a man on the moon and returning him safely to the Earth." One young waitress asked, "Do you think they really went to the moon?" This dark cloud calls into question our nation's willingness, maybe even our ability, to dare great things. It raises disturbing questions: Are America's best days behind us? Will our future be dimmer than our past?

Human spaceflight has been accomplished only by the United States, Russia and most recently China. India has announced its intention to develop such capabilities. Having visited several space facilities in China and India this past year, and met their aerospace engineers, I must say that I am very impressed by the methodical, disciplined approach both countries have taken in developing their space industrial base and capabilities. The national economies of these countries exceed in scale the economy of the United States as it existed in the early 1960s. Thus, if they wish to send their own astronauts into space, it is simply a matter of national will, of choosing to do so. Europe and Japan clearly have the economic and technical wherewithal to do so as well; for either of them, it is again simply a matter of making the strategic choice to do it.

Today is the 46th anniversary of man's first foray into space. That man was a Russian, Yuri Gagarin. Today, a titanium statue of him rises 40 meters above a Moscow square. I will believe that we as a nation truly understand the importance of space to the future of our society when a similarly prominent statue is erected in honor of Alan Shepard, John Glenn or Neil Armstrong.

President Kennedy was the first of our nation's leaders to recognize the importance of U.S. preeminence in space; indeed, it was an electoral issue in 1960, the last time that this has been so. President Kennedy understood the strategic value of space power when he campaigned on the theme of the "missile gap" between the U.S. and the Soviet Union. While we now know that the actual gap was in favor of the United States, the misperception of that time is not the issue; my point is that it mattered. And when Kennedy saw the respect accorded the Soviet Union following Gagarin's flight, he understood as well the strategic value of human spaceflight, and the necessity that the United States be in its vanguard, saying "We go into space because whatever mankind must undertake, free men must fully share."

And human spaceflight is a strategic capability for a nation. To me, Kennedy's appreciation of this matter was similar to the way in which President Theodore Roosevelt recognized the importance of sea power around the turn of the last century as a means to increase the United States' economic, security, diplomatic and cultural influence in the world.

Theodore Roosevelt was a mere 24 years old when his book on the War of 1812 was published in 1882. In it, he wrote that for a state as dependent on sea power as America, it was unthinkable that the nation "rely for defense upon a navy composed partly of antiquated hulks, and partly of new vessels rather more worthless than the old." He went on to say that the United States was rising to world-power status, but it could do so only on the back of a powerful and efficient navy.

As many who work in DOD space understand quite well, there is a direct analogy between many of the operating principles between sea and space power. Roosevelt's work was followed by the influential work of Alfred Thayer Mahan, *The Influence of Sea Power upon History*, published in 1890, which became the bible for the development of sea power by the United States in the 20th century. Mahan also recognized that the United States was rising to world-power status, but could do so only with a powerful navy. According to Historian Paul Kennedy, Mahan "showed the intimate relationships among productive industry, flourishing seaborne commerce, strong national finances and enlightened national purpose." None of these themes has, by itself, any direct connection to U.S. preeminence on the high seas. But none was possible without it.

Mahan's theoretical principles were one thing, but it took President Roosevelt "to turn the theory of Mahan's principles of sea power into effective practice, for the furtherance of American interests and values. No U.S. President did that better." Roosevelt turned Mahan's vision into reality. In an audacious move, President Roosevelt's bold dispatch of the Great White Fleet of 16 modern battleships on a 14-month cruise around the world sent a not-so-subtle message that the United States was an emerging world power capable of projecting its influence where necessary. Roosevelt's experience during the Spanish-American War, when a battleship required over 2 months to steam around Cape Horn from San Francisco to Cuba, prompted him to lead the negotiation for and development of the Panama Canal. The canal continues to be strategically important to our nation even today.

Fifty years ago, first *Sputnik* and then Gagarin sent a similar, and not at all subtle, message about the wherewithal of the Soviet Union. President

Kennedy recognized that this message must be answered with a move even more audacious than that of Roosevelt's Great White Fleet. He recognized that the United States was behind the Soviet Union in human spaceflight, and he recognized its significance concerning the world's perception of leadership, saying: "Those who came before us made certain that this country rode the first waves of the industrial revolution, the first waves of modern invention, and the first wave of nuclear power, and this generation does not intend to founder in the backwash of the coming age of space. We mean to be a part of it—we mean to lead it. For the eyes of the world now look into space, to the moon and to the planets beyond, and we have vowed that we shall not see it governed by a hostile flag of conquest, but by a banner of freedom and peace. We have vowed that we shall not see space filled with weapons of mass destruction, but with instruments of knowledge and understanding. Yet the vows of this nation can only be fulfilled if we in this nation are first, and, therefore, we intend to be first. In short, our leadership in science and in industry, our hopes for peace and security, our obligations to ourselves as well as others, all require us to make this effort, to solve these mysteries, to solve them for the good of all men, and to become the world's leading spacefaring nation."

With President Kennedy's focused goal of "man-moon-decade" in mind, our nation dared to do great things. Webb, Dryden, Seamans, Mueller, Gilruth, von Braun, Kraft, Low, Faget and many, many others were the great leaders of that time. They turned Kennedy's vision into reality and lifted our nation's spirits in the achievement. These men created a lasting legacy and were mentors to thousands of engineers who followed in their footsteps.

Apollo helped create the system engineering discipline that spread throughout our nation's industrial base and found applications in other, diverse fields of the civil and DOD space business, aviation, automotive industry, health care, etc. Like Rickover's nuclear navy, Apollo moved the state-of-the-art forward throughout all of engineering. What is more strategic than that? The need for precise manufacturing methods and engineering standards for human spaceflight systems created a requirement for industry to develop new manufacturing methods and operate to a higher, more precise standard of excellence. The operation of complex, integrated space systems required revolutionary thinking in their development and management. This revolution in our nation's systems engineering discipline was the real spin-off from Apollo, and our nation has benefited immensely from it in many direct and indirect ways.

And while human spaceflight is clearly the most arresting activity any nation can undertake in space, the strategic impact of our efforts in space does not stop there. People seldom recall President Kennedy's breadth of vision, as he also challenged the nation to accelerate the development of communications and weather satellites for worldwide application. Because of that investment, we have a world that is much more connected and safer than otherwise. We have set standards that are followed around the world for the provision of weather data and the distribution of services. And we have greatly extended the goals established in the 1960s. We have two rovers, which have provided a continuous human telepresence on Mars since 2004. We conceived, designed and built the Hubble Space Telescope. We have carried out the first reconnaissance of the solar system, conducted the broadest and most intensive surveys of Earth's weather and climate and developed the first global navigation and communications systems.

So, when we consider the strategic impact of the civil space program, we must ask, what is the value to the United States of pioneering, and leading, enterprises like this, which offer worldwide benefits and lift up human hearts everywhere when we do them? What is the value to the United States of being engaged in projects where we are doing the kinds of things that other nations want to do and including them as partners? I would submit that the highest possible form of national security, well above having better guns and bombs than everyone else, well above "speaking softly and carrying a big stick" as

President Roosevelt suggested, is the security which comes from being a nation which does the kinds of things that make other countries want to join with us to do them. If this is not "strategic," then what is?

I have said many times that I believe that the most important aspect of the International Space Station is the tried and tested partnership that has been forged among the spacefaring nations of Canada, Europe, Japan, Russia and the United States. This partnership has endured tremendous hardships, especially with the loss of the Space Shuttle *Columbia*, and stands by itself as a monumental international accomplishment. The space station partnership has collectively undertaken the largest task ever performed by the civilian agencies of the United States or our international partners; only military coalitions have undertaken larger tasks. With the space shuttle as our primary means for assembling the station, this endeavor rivals the Apollo program in cost and complexity. When completed, the station will be four times larger than the Russian *Mir* space station and five times larger than *Skylab*. It is truly one of the great engineering wonders of the world, akin to such feats as the Great Wall of China; the pyramids of Egypt; the Panama and Suez canals; or the sea walls of Venice.

We can learn from our experience with the ISS and expand on its positive aspects as we move forward. My hope is that by maintaining our commitment to the station, our international partners will view NASA and the United States as good partners through thick and thin, good people with whom to team in future endeavors of space exploration and scientific discovery in exploring the moon, Mars and other worlds. We will also help to drive the creation of a new space industry in low Earth orbit and beyond in such a way that NASA becomes a reliable and supportive customer for that industry. This is the space analogy to Mahan's "flourishing seaborne commerce," and it will be a strategic matter for this century and beyond.

At this stage in the development of our plans for a return to the moon and a lunar outpost, it is important that we at NASA not prescribe roles and responsibilities for future international partnerships. Instead, we have defined a minimalist exploration architecture centered around the Orion and Ares crew- and heavy-lift launch vehicles as the first critical elements, with the hope that international and commercial partners will want to augment these capabilities with their own.

We're already collaborating with other nations on a series of satellite missions to map the resources of the moon, which one day will be mined to help establish a permanent lunar outpost. More than half of NASA's armada of over 50 robotic science missions involve some form of international participation, and almost two-thirds of our science missions on the drawing board today have an international component. One of the main reasons why these discussions for future collaboration in exploring the moon together have been so fruitful is that, despite many trials and tribulations, the United States has shown itself to be a good partner. We need to continue that.

Those who think strategically about geopolitical issues measure a nation's influence on world affairs through four fundamental metrics: economic influence such as the size of a nation's economy and the pattern of its trade relations; military influence such as the ability to deploy army, navy, air and space forces around the world; political influence through diplomacy between countries or in coalitions of nations; and cultural influence with regard to how a country projects its values through various arts, media and language. While some of these influences are easier to measure than others, I think we can see from this discussion that what we do in space contributes to all four of these measures of our nation's influence. What the United States chooses to do in space matters.

"If we can do this, we can do anything."

We could also do nothing. It is a fairly simple choice, really. We could choose to do great things; we could simply sit back and watch; or we could choose to mock those who dare even to try. These are the questions I asked earlier: If our great-great grandparents accepted the challenge of expanding their frontier, will our generation do less? And if so, why? Are America's best days behind us? Will America's future be dimmer than its past?

I have raised these questions, but it is those of you here who must answer them. They are not only strategic choices for our nation, they are also personal choices. All of us, and each of us, must consider the real reasons why we dare to explore this New Frontier.

In conclusion, I would like to leave you with one final thought. Some people have asked me recently about the changes in leadership of Congress and how the next presidential election might change "the Vision." Those questioners are precisely the people who like to be armchair quarterbacks on space policy issues, when what we really need is to focus on the tasks before us and the pace of the work to be done, rather than fomenting discord and putting space policy in partisan, political terms. I would like to echo President Kennedy's advice on the day before he was assassinated, when he spoke in San Antonio, saying: "For more than 3 years I have spoken about the New Frontier. This is not a partisan term, and it is not the exclusive property of Republicans or Democrats. It refers, instead, to this nation's place in history, to the fact that we do stand on the edge of a great new era, filled with both crisis and opportunity, an era to be characterized by achievement and by challenge. It is an era which calls for action and for the best efforts of all those who would test the unknown."

If we can do this, we can do anything. Let's try.

Thank you.

What the Hubble Space Telescope Teaches Us about Ourselves

Michael D. Griffin
Administrator
National Aeronautics and Space Administration

Remarks at the
Institute for Human and Machine Cognition
July 24, 2008

We at NASA deal with the issues surrounding man-machine interfaces everyday, in flying the International Space Station, controlling over 50 Earth and space science missions in operation today, developing new flight control algorithms and avionics for future aircraft, or building the next generation of space vehicles to return Americans to the moon and, later, journeying even deeper into our solar system. To carry out our mission of space exploration, scientific discovery and aeronautics research, we must understand the conditions our machines will face and how they will behave under those conditions because mission success and, indeed, the lives of our astronauts depend upon our machines and the technical acumen of the scientists and engineers who develop and operate them.

I thought it appropriate to speak tonight about the Hubble Space Telescope, one of the greatest machines NASA has ever built, and about our relationship with that machine and what it has taught us about our universe and, more importantly, ourselves. In October, astronauts on Space Shuttle *Atlantis* will rendezvous with Hubble to repair and upgrade it for the fifth time in its nearly two decades of service. When they leave, it will be better than ever. It will be better than anyone ever imagined that it might be back when I was working on the project some 25 years ago.

The story of this scientific and engineering marvel is one of bold vision, imagination and audacious risk-taking, but also perseverance and ingenuity when, as sometimes happens, not all risks are successfully negotiated. It is a story that transcends science, with Hubble images on display today in art museums or in homes where no scientist lives. But we all know that these images are far more than a just a bunch of pretty pictures. Hubble has observed the birth and death of stars not unlike our own solar system. It has shown the collision of the comet Shoemaker-Levy 9 with the planet Jupiter, not unlike the asteroid collision 65 million years ago that wiped out the dinosaurs then roaming Earth. It has peered through a tiny knothole in the night sky, deep into the early universe, finding thousands of galaxies where our own human eyes would see only a patch of darkness. It has found the galaxies in our universe to be accelerating away from each other at a rate faster than any astrophysicist, including Edwin Hubble, ever predicted, allowing new insights into the birth and eventual fate of our universe, while raising new mysteries about dark matter and dark energy, constituents of a universe that, in all humility, astrophysicists must admit we barely understand today. Hubble has become a cultural icon while remaining an instrument of fundamental scientific discovery. It is unique in human history in its ability to occupy a place of prominence in both art museums and scientific journals.

The birth of Hubble, with its launch in April 1990, would not have caused anyone to envision this outcome. Hubble's first images were unaccountably blurry, and analysis of its optical system revealed that a 2.3 micron error had been introduced in the grinding of its 2.4-meter primary mirror. The width of an average human hair is 80 microns, so the error was almost unimaginably small. But as this audience will understand, it is a huge error in terms of the optical wavelengths that a telescope must manipulate if it is to function. This mistake was devastating to the astronomical community. It was equally devastating to NASA's credibility. NASA was the brunt of jokes on late night talk shows, with the Hubble being compared to the *Titanic*, the *Hindenberg*, and the Edsel.

I have said that in the space business we live by a creed of excellence, or die without it. With Hubble, we faced a situation where this small error, left unchecked, called into question our ability to live by that creed. The jokes were cruel, leveling charges that NASA no longer had "The Right Stuff," in Tom Wolfe's elegant phrase. While such talk unfairly denigrates the many dedicated engineers, scientists and technicians who work late into the night to maintain the high standard of most of our endeavors, even the slightest error on such a highly visible project calls into question what happened and, above all else, who was to blame.

Maybe this institute should study this peculiarly human trait—the predilection to "kick those who are down." For me, it always calls to mind President Theodore Roosevelt's great speech, "Citizenship in a Republic," with its famous excerpt about "the man in the arena." Few of those offering criticism of the Hubble mistake were capable of understanding its nature or origin, or indeed anything else of how Hubble was designed, or of the exacting tolerances to which it had to be built or of the tradeoffs that engineers face when deciding how to allocate scarce resources to multiple, competing concerns. As someone who has served on numerous failure boards, and has had to lead teams out of despair, I can only say that criticism from those who are both inept and uninvolved serves no useful function. It cannot even make us feel worse about ourselves than we already do, when we have failed. But it does seem to be a constant companion of bold endeavors, the dark side of human progress. A long career in the space business, with too many opportunities to observe this behavior, has caused me to come to the belief that there is, or at least should be, such a thing as earning the right to hold an opinion.

But I digress. In the aftermath of the Hubble debacle, some Washington policymakers called for an end to NASA altogether. But we don't cast aside human frailty when we venture into space, and wiser heads understood that reaching for the unknown requires the fortitude to deal with adversity.

As President John F. Kennedy warned Congress and our nation in May, 1961, when—with 15 minutes of human spaceflight to our credit—he set forth the challenge to go to the moon, "If we are to go only halfway, or reduce our sights in the face of difficulty, in my judgment it would be better not to go at all."

Thus, the Hubble scientists and engineers set their sights on fixing the telescope. The first step was to characterize precisely the observed error in the primary mirror, and then craft a corrective lens for the aberration. The crew of the first servicing mission to the Hubble trained intensively for one of the most complex shuttle missions ever undertaken, with five spacewalks and over a hundred specialized tools to correct the optics, while also installing new solar arrays, gyroscopes, and other electrical components. They also upgraded the telescope with a new wide field and planetary camera.

You all know today that this first shuttle mission to service the Hubble, as well as the three which followed, were huge successes. The Hubble dazzles us with the splendor of our universe, but during those grim years between 1990 and 1993, its awe-inspiring success was far from certain. If you didn't know the core strength of the NASA team when the chips are down, you might have bet against us. You would have lost.

And that is why, to me, the most meaningful lesson from the Hubble has more to do with our human nature than with any of the secrets of our universe. That is, in the face of adversity, we must resolve to persevere. To that end, I know, because I see it everyday, that NASA still has "The Right Stuff."

Now, I must take a moment to acknowledge those who risked their lives to make the Hubble such a success. Every astronaut I know who has been on a Hubble mission has a special place in his or her heart for that machine. They believe it to be a part of something greater than themselves, that the risk of their lives is worth the promise of unlocking the secrets of our universe for future generations. As David Leckrone, Hubble program's senior scientist, once said:

> We are privileged to be the first generation of *homo sapiens* to gain a clear and deep view of the visible universe. And what we see 'out there' is staggering in its beauty, awesome in its scale and shocking in the way it has upended our preconceived notions about how nature works. You don't have to be a scientist to grasp this. Any thinking person who has come in contact with Hubble images and Hubble discoveries seems to find exhilaration in the notion that our place in the grand scheme of things is now better defined than in all of prior human history.

Dave is so right. And yet, his comment makes a great preface to an observation, which will probably set you back a bit. Science is not everything we do at NASA, nor should it be. And, while the advancement of science is of fundamental importance at NASA, and scientific discovery has a key role in human spaceflight, it is not the most compelling reason to do it.

I would like to take some time to explain why I believe this to be so because numerous critics have called into question the cost and risk of journeys to the moon, Mars, and the near-Earth asteroids or the construction of the space station, which we are using as an engineering test bed to learn how to sustain such journeys. So let me try to provide some food for thought for you tonight. Some of you will disagree with me and thus spark a worthwhile debate. I never learn a thing by talking with people who agree with me.

To me, NASA's manned missions to the Hubble are qualitatively different from our other human spaceflight endeavors. The difference is fundamental and important. And while our other efforts may not seem, today, to be as noble and worthwhile as servicing the Hubble, they are in the long run more important to the future of the human race. Allow me, if you will, to try to explain why I believe this to be so.

Surviving off-planet, in a different environment having different natural resources than those we have come to understand and take for granted, without the ability to drive to the nearest supermarket or doctor's office, is a qualitatively different experience than a brief foray into low Earth orbit. Not many will realize it, but NASA and our international partners have maintained a permanent human foothold in space onboard the space station since October 2000. The hard lessons of living and working in outer space 24/7/365 are much different than those of an intense, 2-week campaign to service a scientific instrument like the Hubble; to deploy a mission like Galileo to Jupiter or Cassini to Saturn; or to conduct other research, as has been done on many individual shuttle missions.

So, when we begin our halting steps back to the moon in the next decade, or a journey to Mars in about 25 years, we will need to know what we must bring with us, but also how we might live off the land with the resources available to us when we arrive. And after we test the hypothesis that we can survive on other worlds, we then need to determine whether such outposts can become economically viable—meaning, is there anything to do there which is worth the investment to do it? Many today will assert, without benefit of proof, that the answer is categorically "yes," while others believe that the answer is "no." In my opinion, no one today can know the answer. The answer can be found only by experiment. In that sense, the purpose of today's human spaceflight program is to conduct such experiments, to explore and develop options, to unveil possibilities for future generations.

This experiment will be conducted in space over the course of the coming centuries by people from Earth. Only the language, culture and motives of the experimenters remain to be determined. I hope that this experiment will always find Americans, in company with our international partners, as first among equals on the frontiers of their time.

The experiment will be not dissimilar to those conducted by our ancestors far removed in space and time, when they left East Africa looking for an easier existence elsewhere. It is not dissimilar to that conducted by our more immediate ancestors, just a few centuries ago, when they began to explore and settle what, to Europeans, was "the New World." In that context, I might note that it required the long-term investment of kingdoms, governments, commercial industry and private citizens for many generations before it could honestly be said that "the New World" provided a positive return on investment for society at large.

And on a smaller scale, our experiment in space will not be dissimilar to that conducted by Thomas Jefferson, when he risked impeachment to consummate the Louisiana Purchase and then sought congressional financing for what became the Lewis and Clark Expedition 200 years ago. By the way, Lewis and Clark overran their budget, lost a considerable amount of their equipment, fell so far behind schedule that they were given up for dead and failed to achieve their primary goal—finding a suitable water route from the headwaters of the Missouri River to the Pacific Ocean. Does anyone here think their effort was wasted?

Venturing into space is similarly an experiment, but one eminently worth conducting, for several reasons. First, I strongly believe that there will be near-term benefits to science, technology, economics and national security as we begin to incorporate the solar system into our sphere of influence, as Science Advisor Jack Marburger framed the issue a few years ago.

We do not need to dwell upon the benefits to human society of scientific advances. We are on the verge of developing a new paradigm, a new view of how the universe is constructed. The last time—a century ago—that such an experience was forced upon us, it was accomplished through the work of Albert Einstein and his elucidation of relativity and quantum mechanics. Today these disciplines underpin much of modern technology, and form the backdrop of physics against which new ideas are interpreted. What will be the implications of forming new theories which embrace the experimental findings that 96 percent of the mass-energy of the universe is comprised of dark energy and dark matter, things we don't yet even pretend to understand?

Regarding technology, what is the benefit to a society which learns how to do what no one else has ever done? No human activity is more demanding, across a broader range of disciplines, than space exploration, nor is there any which produces greater returns from its mastery. Two generations and more ago, in what I consider to be the best speech he ever gave, President Kennedy said, "We choose to go to the moon in this decade and do the other things, not because they are easy, but because they are hard, because that goal will serve to organize and measure the best of our energies and skills. …" As a nation, we are still reaping the benefits of the Apollo investment, but they are coming to an end. America is no longer supreme in the world marketplace, not even in aerospace. It is time to move the goalposts, to define some new "hard things," to move outward again, for precisely the reasons Kennedy articulated so long ago.

A vigorous civil space program offers collateral benefits to national security as well. When I have spoken of this in the past, it has usually elicited some surprise. But I think those who are surprised are taking too narrow a view of "national security." For the last century, the United States has been a world power, even if at times we did not aspire to or even recognize that fact. As such, we have assumed certain responsibilities for leadership on the world stage, and in that capacity it is inevitable that we have been, and will be again, called upon to make decisions and take actions that displease other nations and societies. We cannot possibly please everyone, and we cannot retire from world affairs.

But it is equally true that we cannot prosper if every hand is against us. So if we must do hard things, it behooves us also to undertake activities which easily attract allies and partners, things which bind us to others in the world community. No activity has shown itself to be of greater inherent interest and excitement to others than has the exploration and development of the space frontier. And so I ask, concerning national security, what is the value of being a nation, a society, which leads the world in an endeavor that excites all others, one in which every nation that can do so seeks to partner with us?

These are some of the specific benefits I see accruing to the nation which leads in the exploration of space. But I also believe that, in the long term, it will be important for the survival of *homo sapiens* to inhabit planets other than Earth. It will be in our interest to develop the technical capabilities to avoid the many cosmic collisions which we have now documented in the geological record. The comet Shoemaker-Levy 9 consisted of at least 21 discernable fragments with diameters up to 2 kilometers wide. Even one such collision with Earth would be devastating, and it doesn't have to be a dinosaur-killer. An impact like the Tunguska event of 1908 could destroy the cultural and economic fabric of a nation, should it land in a populated area instead of the Siberian wilderness.

And so I believe that long-term survival, scientific discovery, economic benefit and recognized leadership in great endeavors provide a worthwhile rationale for sustaining our nation's human spaceflight efforts. This and our endeavors in robotic Earth and space science, and our work in advanced aeronautics, are purchased with an investment in NASA of less than 0.6 percent of the federal budget of the United States. (If any of you happen to be an average Americans, this figure will surprise you, as polls reveal that the 50-percentile American believes that NASA receives over 24 percent of the federal budget, comparable to that of DOD.)

My view is that our efforts in human spaceflight are, in actuality, far more meaningful than the "flags and footprints" rationale with which critics of human spaceflight like to denigrate Apollo or future voyages to the moon and Mars. Survival, leadership in great enterprises and economic benefit are real and acceptable reasons why humans should continue to explore space beyond what robotic spacecraft can achieve.

Throughout mankind's time on this world, we have gazed up at the night sky and attempted to make sense of the stars, planets, comets and asteroids, speculating about what they might mean. While we are lucky enough to be the first generation to see the universe with the clarity Hubble offers, I firmly believe that we also need to journey beyond "the surly bonds of Earth," in order to see the universe with our own eyes. In the words of poet T.S. Eliot, "Only those who will risk going too far can possibly find out how far one can go." Expanding the range and scope of human action is a goal fully as noble as that of scientific discovery.

In our hearts, we know these things. We know that space is the frontier of tomorrow, and that the frontier can only be ours with "boots on the ground." We know from even the most casual reading of history that nations, which shrink from the frontiers of their time, shrink also in their influence on the world stage. We know these things, and yet we also see that Americans today do not feel the urgency for preeminence on the space frontier that we felt in the 1950s and 1960s. Sometimes I wonder if we are a bit tired or distracted from other, urgent crises to recognize what that preeminence means for America.

And so I am reminded of Edgar Allan Poe's "gallant knight" in search of Eldorado and who, in his fatigue, asks a "pilgrim shadow" where it might be. "Over the Mountains of the Moon, down the Valley of the Shadow, ride, boldly ride," the shade replied— "If you seek for Eldorado!"

Sometimes, there is no rest for the weary.

Thank you.

Bob Peters, of Kendel Welding and Fabrication, welds part of an internal access support for the Ares I-X upper stage simulator at NASA's Glenn Research Center in Cleveland. In 2009, NASA will launch Ares I-X, a test flight for the new Ares I rocket. Ares I will launch astronauts on missions to the space station, the moon and beyond. (21995main_image-eng)

This artist's concept illustrates how astronomers have determined the distance to an invisible Milky Way object called OGLE-2005-SMC-001 using a depth-perceiving trick called parallax. Different views from both Spitzer and telescopes on Earth are combined to give depth perception. (DARK_BODIES)

With its thermal-infared sensor pod under its left wing, NASA's Ikhana unmanned aircraft cruises over California during the Western States Fire Mission. (ED07-0186-13)

The X-48B Blended Wing Body research aircraft banks smartly during the Block 2 flight phase. (ED08-0092-03)

A drafting room at the NACA Airplane Engine Research Laboratory (AERL), now known as the NASA Langley Research Center, Hampton, Virginia. (GPN-2000-001839)

(March 24, 2008) — The STS-123 crew used part of its last full day onboard the space station taking in-space crew portraits. These six astronauts launched aboard the Space Shuttle *Endeavour* on March 11 and are scheduled to return aboard it on March 26. Shown are astronauts Dominic Gorie (top center), commander; Gregory H. Johnson (bottom center), pilot; along with astronauts Rick Linnehan (top left), Mike Foreman (top right), Robert L. Behnken (bottom left) and JAXA's Takao Doi (bottom right), all mission specialists. (ISS016-E-033709)

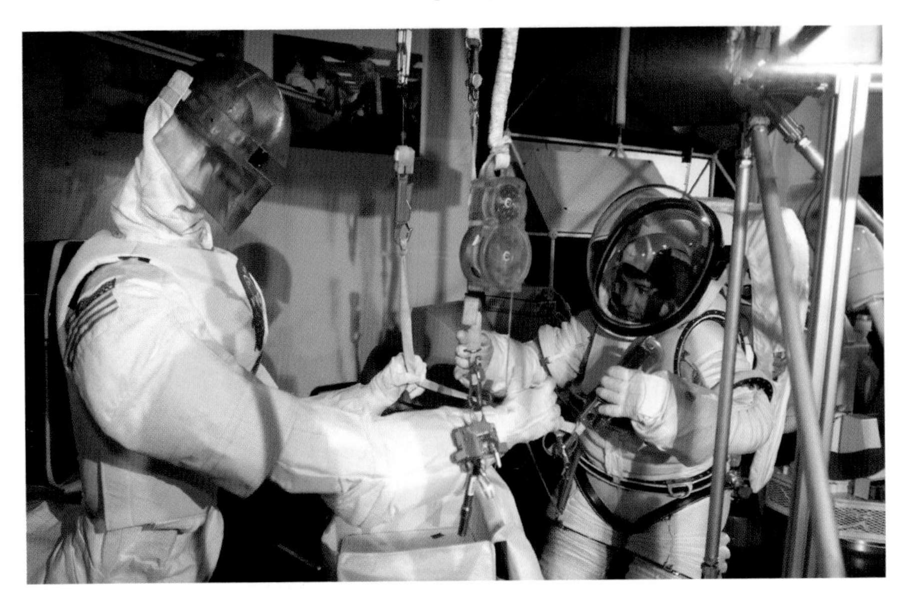

Bob Peters, of Kendel Welding and Fabrication, welds part of an internal access support for the Ares I-X upper stage simulator at NASA's Glenn Research Center in Cleveland. In 2009, NASA will launch Ares I-X, a test flight for the new Ares I rocket. Ares I will launch astronauts on missions to the space station, the moon and beyond. (21995main_image-eng)

This montage of images of the Saturnian system was prepared from an assemblage of images taken by the Voyager 1 spacecraft during its Saturn encounter in November 1980. This artist's view shows Dione in the forefront; Saturn rising behind; Tethys and Mimas fading in the distance to the right; Enceladus and Rhea off Saturn's rings to the left; and Titan in its distant orbit at the top. (PIA01482)

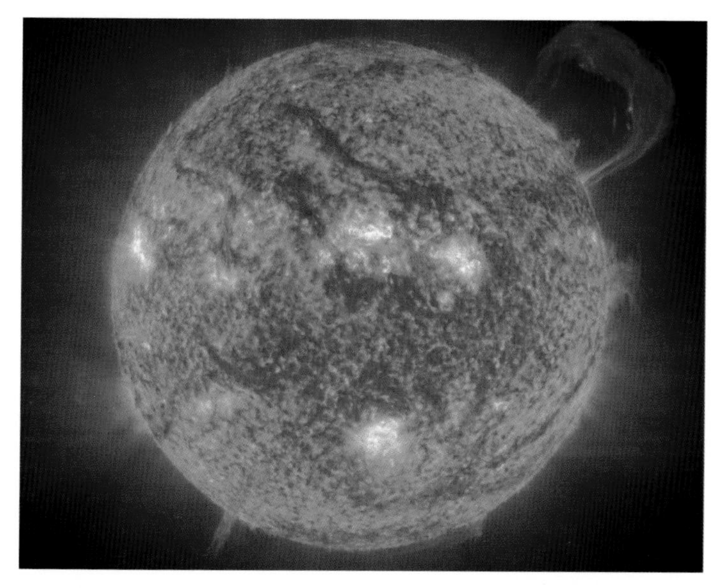

An Extreme Ultraviolet Imaging Telescope (EIT) image of a huge, handle-shaped prominence taken on September 14, 1999 in the 304 angstrom wavelength. Emission in this spectral line shows the upper chromosphere at a temperature of about 60,000 degrees kelvin. Every image feature traces magnetic field structure. The hottest areas appear almost white, while the darker red areas indicate cooler temperatures. (PIA03149)

A montage of New Horizons images of Jupiter and its volcanic moon Io, taken in early 2007. The Jupiter image is an infrared color composite taken by the spacecraft's linear etalon imaging spectral array (LEISA) at 1:40 UT on February 28, 2007. Blue denotes high-altitude clouds and hazes; red indicates deeper clouds. The bluish-white oval is the Great Red Spot. The observation was made at a solar phase angle of 75 degrees. The Io image was taken at 00:25 UT on March 1, 2007 by the panchromatic long-range reconnaissance imager (LORRI), with color provided by the multispectral visible imaging camera (MVIC). (PIA10102)

An artist concept illustrating the new view of the Milky Way along with other findings presented at the 212th American Astronomical Society meeting in St. Louis, Missouri. The galaxy's two major arms (Scutum-Centaurus and Perseus) can be seen attached to the ends of a thick central bar, while the two minor arms (Norma and Sagittarius) are less distinct between the major arms. The major arms consist of the highest densities of young and old stars; the minor arms are primarily filled with gas and pockets of star-forming activity. (PIA10748)

(December 12, 2006) — Backdropped by New Zealand and Cook Strait in the Pacific Ocean, astronaut Robert L. Curbeam Jr. (left) and ESA astronaut Christer Fuglesang, both STS-116 mission specialists, participate in the mission's first of three planned EVA sessions while construction continues on the space station. (S116-E-05983)

(June 11, 2008) — Backdropped by a blue and white part of Earth, the space station is seen from Space Shuttle *Discovery* as the two spacecraft separate. Undocking of the two spacecraft occurred at 6:42 a.m. CDT on June 11, 2008. (S124-E-009982)

(February 7, 2008) — The Space Shuttle *Atlantis* and its seven-member STS-122 crew head toward Earth-orbit and a scheduled link-up with the space station. Liftoff from Kennedy Space Center's launch pad 39A occurred at 2:45 p.m. EST. The crew's prime objective is to attach the Columbus laboratory to the *Harmony* module. Onboard are astronauts Steve Frick, commander; Alan Poindexter, pilot; Leland Melvin, Rex Walheim, ESA's Hans Schlegel, Stanley Love and ESA's Leopold Eyharts, all mission specialists. (STS122-S-051)

Young stars want to show their independence. This Spitzer view shows a stellar version of the "terrible twos". The stars are beginning to move away from their formative cloud, seen in red and green. Jets come off the young stars as they make their way into the cosmos. (TerribleTwos)

The Eagle Nebula (M16) showing a portion of a pillar of gas and dust. Light from nearby bright, hot and young stars sculpts the cloud into intricate forms, causing the gas to glow. (Top_Pillar M16)

MODIS image of Hurricane Katrina taken on August 28, 2005. Hurricane Katrina strengthened into a powerful Category 5 hurricane overnight with sustained winds of 160 mph. The National Hurricane Center put out a special advisory on the hurricane's gain in strength just before 8 a.m. EDT. The boost came just hours after Katrina reached Category 4, with wind of 145 mph, as it gathered energy from the warm water of the Gulf of Mexico. (Credit: NASA/Jeff Schmaltz, MODIS Land Rapid Response Team)

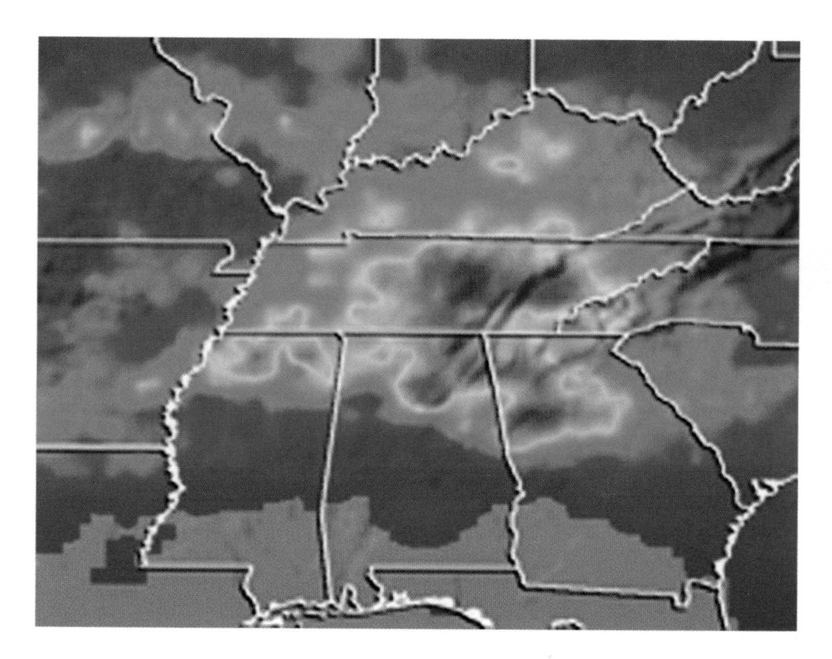

Torrential rainfall from a 2003 storm in the Southeast resulted in massive accumulations of rain (red). Similar data from NASA's Tropical Rainfall Measuring Mission (TRMM) satellite has revealed that more rain falls midweek. (Credit: NASA)

This mosaic of two Cassini images shows two of Saturn's moons, Pan and Prometheus, creating features in nearby rings. The images were taken in visible light with the Cassini spacecraft narrow-angle camera on August 15, 2008. The Cassini-Huygens mission is a cooperative project of NASA, the European Space Agency and the Italian Space Agency. (Image Credit: NASA/JPL/Space Science Institute)

James Webb Space Telescope full-scale model on display at The International Society for Optical Engineering's (SPIE) week-long Astronomical Telescopes and Instrumentations conference, May 25–30, 2006. The model built by the prime contractor, Northrop Grumman, is constructed mainly of aluminum and steel, weighs 12,000 pounds, and is approximately 80 feet long, 40 feet wide and 40 feet tall.

Part 2.

NASA, Science, Commerce and Engineering

The Rocket Team

Michael D. Griffin
Administrator

National Aeronautics and Space Administration
The 19th Annual Von Braun Dinner
Huntsville, Alabama

October 22, 2007

Space exploration is a complex story, a rich story, full of drama and despair, pride and pathos. It is a story we need to tell our children and grandchildren, lest they forget why we explore what John F. Kennedy referred to as the "New Frontier" of space. There are many distractions in modern life; and I believe it is necessary for us to discuss openly with the public the principles that led us as a nation to embrace space exploration five decades ago.

I recently read an interview with Actor Bill Pullman, who is famous among those of us who watch science fiction movies for being the president who beat the aliens in the movie *Independence Day*. Pullman wrote and produced a new play about the International Space Station Expedition 6 crew, Ken Bowersox, Don Pettit, Nikolai Budarin and their trials and tribulations aboard the station following the loss of the Space Shuttle *Columbia*. When Pullman was asked the question about how he first learned of the *Columbia*'s loss, he responded quite simply and insightfully: "It was a Saturday morning, and I think I was in the car driving. I had to go run to get milk and I heard the radio report. I remember pulling off to the side of the road and listening to it. It was stunning to me. I was aware that I hadn't kept up on it. I wasn't somebody who was aware that they had even gone up. Suddenly I became hugely interested."

Probably everyone in this room remembers that Saturday morning of February 1, 2003. I know Dave King does. He spent the next several months in Texas and Louisiana leading the debris recovery and investigation efforts for *Columbia*. Many of us that morning were probably going about our lives in a manner similar to Bill Pullman when we were pulled in by the television or radio with the news. We called and e-mailed our family and friends in the space business, and most importantly, we rolled up our sleeves and went to work finding the cause of the accident, fixing it and continuing the journey.

There are galvanizing moments in our lives, moments we remember forever, moments when we hold our breath in the realization that the events unfolding before us will forever change the course of our lives. These are the events for which we remember precisely where we were and what we were doing, what we saw and what we felt when we first heard. For the rest of our lives, we return to them in quiet introspection, thinking about how the world changed in those moments. Those who are older can recall many such. For the oldest among us, there is still Pearl Harbor and later Hiroshima. For those a bit younger, the assassinations of John and Robert Kennedy and Martin Luther King, Jr. might be the first. Younger still, there is the fall of the Berlin Wall or September 11th. Not one of us ever, ever forgets such things. And for those here tonight, there are many more such milestones even closer to home: Sputnik,

Yuri Gagarin, John Glenn, the Apollo Fire, "The Eagle has landed" and "one small step," *Challenger* and *Columbia.*

These are the things, too often crises, which shape the course of human events. Thus, tonight I will pose for you a question, which I hope will stir some debate: Why does it take a crisis to capture our attention?

This is a simple question without a simple answer. However, I do believe that it is fundamental to some of the problems we face in explaining the importance of space exploration to the American public or to our children. There are many distractions in our lives, distractions that make it difficult to distinguish between what is urgent and what is important. It is easy to become complacent about or even apathetic toward the signals that, too often in the clarity of hindsight, show that another crisis looms, that action should be taken.

Crises can take many and various forms, and always—always—in the investigation that occurs after an accident or a tragedy, we find that there were warning signs, that there were people who connected the dots but were not heeded. Churchill was right about Hitler years before Hitler proved him so. And as Admiral Gehman said of the Space Shuttle *Columbia*, "The machine was talking; but why was nobody hearing; how were the signals missed?"

Even worse, with the passage of time we seem to forget the lessons learned from those crises that occurred many years ago. Time heals the wounds, the fear and the pain we felt when the galvanizing moment occurred. We move on. And slowly, our complacency grows back. The great engineering Educator and Author Henry Petroski writes about this facet of humans and their organizations in his book, *Design Paradigms: Case Histories of Error and Judgment in Engineering*. He cites a trend, two centuries old, of major bridge disasters occurring about every three decades. Younger engineers who did not experience the community-wide trauma of such an event do not fear it, do not believe it can happen to them and do not embrace its lessons as deeply as those who were there. The cycle thus begins anew.

As managers, we must understand this aspect of human nature and fight against it. We must inspire and reward perseverance and persistence to the task before us. We must check, re-check and check again to hear what our machines and our people are saying. All of us—from assembly-line technicians through young and mid-career engineers to center directors and associate administrators within NASA—have the responsibility to speak up if we believe that something is amiss with our part of the complex machine. Other people may disagree with any given concern or may simply see things differently; in fact, it is guaranteed that they will. And no decision can be made that doesn't leave at least one group feeling as if their concern has been set aside. But it is still everyone's responsibility to offer their own judgment on a controversial issue. The final decision cannot be made better by the lack of debate. In this way, sometimes a crisis can be averted.

This takes me to last week's Flight Readiness Review for STS-120. We should all applaud the folks from NASA Engineering and Safety Center (NESC) who brought forward their concerns with regard to the integrity of the Space Shuttle *Discovery* wing leading edge. We have a new inspection technique that, if nothing else, demonstrates that we don't know as much about the reinforced carbon-carbon (RCC) panels that comprise the wing leading edge as we thought we did. This realization brings with it the concern as to whether several of the panels had adequate margin for flight. We had a good, healthy engineering discussion, culminating in a majority, but not unanimous, decision that we have an acceptable level of risk to launch the space shuttle. The bottom line is: I don't think we're seeing new behavior in the RCC panels. I think we're seeing how the panels we've always flown look, when inspected via a new technique. But I will say here that I simply could not be happier with the manner in which the NESC folks pursued, and brought forth, this concern.

In the space business, we live up to a creed of excellence or die from the lack of it, and we make our entire society better for the acceptance of that challenge. We are not perfect; we do not have perfect knowledge of our machines or the environment in which they will be operating. Our machines are no more perfect than we ourselves. A quote that I love goes like this: "Excellence is the result of caring more than others think wise, risking more than others think safe, dreaming more than others think practical and expecting more than others think possible." My hope is that we inspire our people to work—and work hard—toward the goals of the missions placed before us, as our forebears did. That's what it takes. This is rocket science.

NASA is a high-performance organization, working on large, complex engineering systems on their way to Mercury, Mars, Pluto, and with the Dawn mission, the asteroid belt between Mars and Jupiter. Weather permitting, my hope is that tomorrow or later this week we will launch Space Shuttle *Discovery*, commanded by Colonel Pam Melroy, to the station on the STS-120 mission. This shuttle mission will deliver the Italian-built *Harmony* Module to connect the European and Japanese laboratory modules that will be flown on the next two missions.

But, we will only launch after checking, re-checking and checking again. Tonight, as I speak, hundreds of technicians and flight controllers are working toward that launch. Tonight, here at Marshall, payload operators are working on experiments onboard the station. Tonight, the Expedition 16 crew commanded by Peggy Whitson will soon wake to begin preparing for *Discovery*'s arrival. On November 2, we will celebrate 7 years of continuous manned spaceflight operations aboard the space station. Many, many people said that such a goal could never be reached, but as Meriwether Lewis wrote in his journal, "we continued on."

In discussing the great things we have accomplished and seek yet to do, I need to return to my original question: why does it take a crisis to get the

American people's attention? It is frustrating sometimes for those of us in the space business to realize that many people in the American public are not aware, or do not care, about the things we are accomplishing, often for the first time in history.

We saw this for the first time with the lunar missions that followed Apollo 11, except most famously the harrowing Apollo 13. Some people lost interest—lost interest!—in seeing a precious few of their fellow Americans begin the exploration of an entirely new and unknown world. Today, it can be frustrating when some young people actually question whether we ever really landed on the moon. However, it has been almost 35 years, and enough time has passed that many Americans forget the importance of these events in their time. In a way, it's a lot like Petroski's observations concerning the three-decade cycle in major bridge collapses. New generations sometimes need to relearn the lessons so painfully gathered by their fathers.

Perhaps, that is what prompted the *Columbia* Accident Investigation Board to observe: "The U.S. civilian space effort has moved forward for more than 30 years without a guiding vision." That was a damning statement, citing (as it did) a lack of leadership in space policy, a strategic interest for the United States reaching to the highest levels of our nation for over a generation.

Earlier this year, Mike Coats asked me to speak at a dinner in Houston. It was "budget season" in D.C., and I didn't have time to write a speech, or even to seek help from any of my colleagues, who might have been willing to furnish a draft for editing. I was simply out of time when the dinner arrived, and so I stood up to speak with nothing more than the benefit of a few notes I made on a napkin during dinner. Thus, I spoke more from the heart and less from my analytical side than is customary for me. I discussed the "real reasons"—as compared to the "acceptable reasons"—why those of us in the space business make the sacrifices we do to pursue the dream and the challenge of spaceflight. Some of you may have been there or perhaps have read the speech, which later appeared in *Air and Space* magazine. I've been enormously surprised by the outpouring of positive feedback I've received in regard to that speech, far more than for any other speech I have ever given. With those thoughts, I must have touched a sensitive nerve that the analytical side of my brain did not know was there. The real reasons which drive those of us who are in this business are, I think, more visceral, or even spiritual, than can be expressed by means of any tangible rationale.

While NASA's budget is about half a cent out of every federal budget dollar, spaceflight in all its forms is a strategic capability for this nation. We must understand the real reasons why that is so, we must explain those values to our children, to their children, to the public and to the nation's leadership, lest it just slip away.

Thus, maybe the reasons why the American public is not aware of what we're doing in space, of what we're trying to accomplish, is that we're not explaining it well enough. Maybe the scientists and engineers in this room need the help of folks like Miles O'Brien, Neil Tyson, Homer Hickam, Tom Hanks, Bill Pullman and many, many others who are far more charismatic than I will ever be and who know how to weave the fabric of such a story. For those of us in the space business, this is our story, a complex story full of richness, daring, drama, comedy and pathos. While I don't pretend to know all the different ways to tell it, or maybe any of them, I do know it cannot be condensed down to a bumper sticker slogan. But it can be distilled. "This cause of exploration and discovery is not an option we choose," as President Bush put it in his eulogy to the *Columbia* astronauts. "It is a desire written in the human heart. We are that part of creation which seeks to understand all creation. We find the best among us, send them forth into unmapped darkness and pray they will return. They go in peace for all mankind, and all mankind is in their debt."

A few weeks ago, many television news shows and newspapers recognized the 50th anniversary of the launch of the first man-made satellite, the Russian

Sputnik. Some commentators noted the galvanizing reaction of this event on the American public and our national leadership around the question of whether we were falling behind in recognized leadership in the world, falling behind the Soviet Union in technological competitiveness and how this reaction was primarily a media-driven frenzy. That is the power of the American media then as now. America at the forefront of the frontier is a concept deeply embedded in our national psyche. People who tell stories for a living know this better than I do. Space was the New Frontier, as the junior senator from Massachusetts and future president would say. He was the first of our national leaders to recognize the strategic importance of the new medium, the new arena of space.

President John F. Kennedy also understood what it meant for nations to ignore the telltale signs of a looming crisis, failing to connect the dots. His thesis at Harvard in 1940, *Why England Slept*, compared the failure of the British government to take steps to prevent the rise of Nazism in Europe with allusions to how America was also ignoring its own looming crisis and could be pulled into another world war. Like Churchill, Kennedy spoke up about his concerns, just as I have asked every NASA employee to speak up if they have concerns. In Kennedy's case, when he spoke in his famous speech on May 25, 1961 about the need to "take longer strides," Congress and the American people listened.

In my own small way, I have recently given vent to my thought that the pace of China's space program may be faster than our own. Later this week, China plans to launch its first satellite to the moon. I also believe that, if they so choose, the Chinese have the economic and technical wherewithal to send their *taikonauts* to the moon before the United States plans to return our own. If this happens, we in the United States will not like our position in the world of that time. I am speaking out now because I hope to avert the situation our nation faced 50 years ago with the launch of *Sputnik*.

Even at the age of 8, I was as attuned to events following the launch of *Sputnik* as closely as was possible by watching television and reading *The Baltimore Sun*. The newspapers were full of both soul-searching analysis and rampant second-guessing. We questioned our military plans, our civilian research programs and our educational systems and made changes in all those areas and more. America's readiness—or more properly our lack of readiness—to explore and exploit the space frontier decided a presidential election. *Sputnik* changed everything.

I was in Russia a few weeks ago toasting the 50th anniversary of this accomplishment with my Russian counterpart, Anatoly Perminov. Times have changed. NASA is now paying the Russian Space Agency several hundred million dollars over the next several years for the *Soyuz* and *Progress* vehicles necessary to support the space station. Partly for that reason, we need the help of the rocket team here at Marshall and our industry partners to develop the next-generation Ares rockets as expeditiously as possible. I would rather we spent NASA's funds within the American space industry, first with U.S. commercial systems to support the station, and then the Orion crew vehicle and Ares rockets. This is both important and urgent, and we need to work with the same sense of purpose as our forbears to build these new systems.

While we engineers like to talk about the machines that propel us into space, in a democracy, it is really the American people who ignite our nation's space program. Fortunately for those who care about space, one of the most charismatic men in history was the first Director of Marshall Wernher von Braun, whose memory we honor here tonight. Chris Scolese brought to my attention a wonderful book, *The Rocket Team*, about the life and times of von Braun and the team he built. I commend it to your attention.

There's no need for me to recount to this group the story of the von Braun team and how they built the V-2, Redstone, Jupiter and, finally, the Saturn V. Many of you know far more than I do about these accomplishments.

Von Braun's charisma, technical acumen and leadership in the field of space exploration are legendary.

But do you know what Huntsville was like before von Braun settled here to work in the spring of 1950? The population of Huntsville was 16,000, and the city fathers proudly advertised it as "The Watercress Capital of the World." Von Braun changed Huntsville, the nation and the world in the course of his pioneering efforts in space exploration. Von Braun and other legendary engineers and managers like Glynn Lunney, George Mueller, George Low and Chris Kraft turned President Kennedy's vision into a reality. I've said before, and will do so again here, that James Webb was NASA's greatest administrator for the manner in which he kept those people and their programs pointed in the right direction during the 1960s. Today, young engineers in this audience are following in their footsteps, and pursuing a vision for space exploration which, I hope will be sustainable over the next 50 years.

Look around the room. You are the people whose accomplishments future NASA administrators will toast 50 years from now. You are the ones who will be building the Ares I crew launch vehicle and the Ares V heavy-lift launch vehicle to propel our nation back to the moon. But it will only happen if we all work just as hard as they did. *You* are the new Rocket Team. But not only must we be able to build rockets, we must also re-ignite the passion for space exploration that von Braun conveyed to his team and the nation. This is now our story to tell.

Only a few months before he died, von Braun wrote the following: "While the members of this magnificent team changed with time, the fundamental characteristics of the team itself never did. It always has been characterized by enthusiasm, professionalism, skill, imagination, a sense of perfectionism and dedication to rocketry and space exploration. How can the story of such people and of the exciting programs with which they are involved ever end?"

So, let us resolve that it will not—not ever—end.

NASA's Direction

Michael D. Griffin
Administrator
National Aeronautics and Space Administration
Remarks at the Mars Society Convention
August 3, 2006

This morning, I want to look at NASA's overall mission, rather than at the single goal of a voyage to Mars.

Many know that I am an admirer of NASA's greatest Administrator, James Webb. Webb once characterized his role during the Apollo program in the following way: "The process of management became that of fusing at many levels a large number of forces, some countervailing, into a cohesive, but essentially unstable whole, and keeping it in the desired direction." This is it, exactly, and that perspective serves me well today. There are many disparate goals held by NASA's numerous stakeholders, and we try—very hard—to move the agency forward in a manner which promotes unity among, rather than division between, these stakeholders. It is not easy.

If the blunt truth be told, prior to the loss of the Space Shuttle *Columbia* a few years ago, NASA suffered from a long period of benign neglect by both the public and our government stakeholders concerning the broader purposes of our nation's space enterprise, especially human spaceflight. NASA's last mission to the moon was in 1972, and our nation, as a matter of policy at the highest levels, had chosen to confine itself to low Earth orbit ever since. As I have said on many, many previous occasions, I believe this to have been a crucial strategic mistake for our nation.

We have come a long way since the dark days of the *Columbia* accident in building a consensus as to what goals are worthy of our nation's civil space program. After a lengthy national discourse, the bold challenge of the President's Vision for Space Exploration was endorsed by Congress in the NASA

Authorization Act of 2005. I firmly believe that the Vision is the proper lasting legacy for the astronauts who perished in the *Columbia* accident. It sets a course in space for our generation and, indeed, future generations. The law charges the NASA administrator to "establish a program to develop a sustained human presence on the moon, including a robust precursor program, to promote exploration, science, commerce and United States preeminence in space, as a stepping-stone to future exploration of Mars and other destinations." This is wonderful direction. It tells NASA to make the United States, once again, a spacefaring nation. Nothing more is necessary, and nothing less is appropriate.

The law charges NASA to carryout certain specific programs with the following milestones:

- Return Americans to the moon no later than 2020.
- Launch the Crew Exploration Vehicle as close to 2010 as possible, and not later than 2014.
- Increase knowledge of the impacts of long duration stays in space on the human body using the most appropriate facilities available, including the International Space Station.
- Enable humans to land on and return from Mars and other destinations on a timetable that is technically and fiscally possible.

With this, NASA's stakeholders at the White House and Congress have provided clear direction on the policies and programs that the agency must carryout. And so, while some of you might wish it to be otherwise, NASA's strategic goals are neither solely nor initially focused upon Mars. We are charged with carrying out a broad portfolio of missions in space exploration, scientific discovery and aeronautics research. With the resources projected to be available to NASA over the next 5 years, properly balanced with our other national priorities of Earth and space science as well as aeronautics research,

NASA is on course to complete the space station by 2010 and to bring the CEV online no later than 2014.

This is a national imperative. With the retirement of the space shuttle by 2010, a far more urgent concern to me than a future mission to Mars is taking steps to ensure that we have a smooth transition from the space shuttle to the CEV, and to commercial cargo and crew transport services to the space station. Thus, we plan to award a contract for the design and development of the CEV next month, and in the coming weeks ahead, we hope to enter into Space Act agreements with commercial firms to demonstrate space station resupply capabilities. As I have said previously, if cost-effective commercial services are demonstrated to support the station, NASA will welcome and use them.

So, let me be clear about what is at the forefront of our attention at NASA. The greatest management challenge we face over the next several years is the safe and effective transition to the new exploration systems by completing the remaining space shuttle missions for the assembly of space station, followed by retirement of the space shuttle in 2010 to allow greater focus on missions beyond low Earth orbit. We now have a clear goal for the future. Our current activities can and must be focused to advance that goal. The remaining space station assembly missions and space shuttle flights will allow us to advance knowledge and help train the next generations of space explorers.

And, more broadly, I believe that the most important aspect of the space station is the tried and tested partnership that has been forged among the spacefaring nations of Canada, Europe, Japan, Russia and the United States. This partnership has endured tremendous hardships and stands by itself as a monumental international accomplishment. We can learn from this experience, and expand on its positive aspects as we move forward to the moon and Mars.

At this stage in the development of our plans, it is important that NASA not prescribe roles and responsibilities for future international partnerships. Instead, we have defined a minimalist exploration architecture with the CEV, the crew- and heavy-lift launch vehicles and a lunar lander as the first critical elements, with the hope that international and commercial partners will want to augment these capabilities with their own. We're already collaborating with other nations on a series of satellite missions to map the resources of the moon, and I hope that we'll collaborate on even more missions to the moon and Mars.

This year, we've hosted workshops in order to discuss with our international partners, scientific communities and commercial interests what each of us might do, and what we might do together, in exploring and utilizing the moon. I hope to issue a plan later this fall based on the feedback from those workshops. One of the main reasons why these discussions for future collaborations in exploiting the moon have been so fruitful is that, despite many trials and tribulations, the United States has shown itself to be a good partner on the space station. We need to continue that.

Also this year, we've celebrated the 30th anniversary of the Viking mission and begun to lay the groundwork for the robotic missions to Mars in the next decade, building on the results from the Mars Rovers and Mars Reconnaissance Orbiter. Next year, I hope to make plans as to how to carryout manned missions to Mars, building on the heavy-lift launch vehicles, landers and other capabilities from the lunar exploration architecture. We will especially call for the support of our international partners for this long-term endeavor, as we build on the relationships forged in assembling the space station.

This morning, onboard the station, American astronaut Jeff Williams and German astronaut Thomas Reiter are preparing for a 6-hour spacewalk to install equipment and experiments around the outside of the station. This spacewalk is being choreographed by Russian Commander Pavel Vinogradev who will remain onboard the space station and be televised by cameras on the Canadarm-2. Today's spacewalk exemplifies how nations work together at their best. We make it look easy … perhaps even too easy, to those who are not steeped in the risks of the space business. However, this spacewalk is not simply science fiction. The space station is our nation's most technically challenging project, and this spacewalk is one small step forward toward exploration.

The station is a scientific and engineering test bed that we must use before embarking on long journeys to Mars. In fact, the NASA Authorization Act designates the U.S. segment of the space station as a national laboratory, and I am actively seeking partnerships with other federal agencies like NSF, NIH and DOD, and commercial entities who would use the station for their own experiments while NASA focuses its research directly on exploration-related missions.

For example, we know that a crew on such a long journey to Mars will need a great deal of self-sufficiency as we break the apron-strings of Earth. A three-person crew onboard the space station requires approximately 5,000 pounds of water each year. However, by developing and testing environment and water recycling capabilities for the station, the future logistics resupply required will be significantly less. This will have far-reaching implications for future Mars missions. Likewise, learning to use resources on the moon and Mars must prove far more economical than bringing those resources with us on the journey.

In situ resource utilization, maintenance and logistics, international partnerships and the leveraging of commercial space capabilities will help sustain this Vision for Space Exploration, and they are all being tested onboard the space station.

The priorities for NASA are clear. In the wake of the *Columbia* tragedy, our national leadership realized that human spaceflight is today one of those strategic capabilities that define a nation as a superpower. And I must ask each of you this: Is it possible to envision a future in which America is considered to be a leader in the world, if others can and do conduct exploration and research on the moon, Mars and beyond, and we cannot? International cooperation, leavened by a healthy dose of competition, is what makes America the greatest country in the world. The ultimate goal of the Vision for Space Exploration is not to impress others, or even merely to explore the moon or Mars, but rather to advance U.S. scientific, security and economic interests through leadership in the grandest expression of human imagination of which we can conceive.

We are turning science fiction into reality. We do what others dream.

Science and NASA

Michael D. Griffin
Administrator
National Aeronautics and Space Administration
Goddard Space Flight Center

September 12, 2006

I'd like to share with you my thoughts on some issues that seem to have disturbed quite a few members of the scientific community. Many of my friends have spoken to me about these issues. And there's been a lot of hyperbole flung about in the media during the past several weeks about NASA's "decimated science program," how NASA has rejected its responsibilities in the study of Earth science and how we're not listening to our advisory committee.

So, let me be clear at the outset with my response to such hyperbole: Nothing could be farther from the truth. And frankly, as someone who's spent a good part of my engineering career building NASA science satellites, I think that this sort of unfounded rhetoric hurts the overall space program, including space science. But allow me an opportunity to offer a few points as I see them to ground the discussion in fact before opening it up to your questions. My intent is to change this debate into a more thoughtful, objective dialogue about the issues facing NASA's science and exploration programs than what has been presented in many circles. I'd really like to reduce some of the angst in the community.

As I see it, by any objective measure science is doing well at NASA. Within the context of a national policy mandating a return of humans to deep space and adherence to our international commitment to use the space shuttle to complete the assembly of the International Space Station, NASA is maintaining many vigorous science programs, not the least of which will be the opportunity to reconstitute a productive program of human and robotic exploration of the moon.

The Science Mission Directorate (SMD) fiscal year 2007 (FY07) budget request is $5.33 billion, up from FY06 by 1.5 percent. And we have an administration (not just NASA) that is committed to preservation of SMD funding in FY08–10, albeit at a lower growth rate, 1 percent, than we all would like. In FY11 and beyond, SMD funding tracks agency top line growth.

I must note here that fiscal years 2008–2010 are very, very difficult budget years at NASA, because we are engaged in completing the space station, while at the same time trying to gain ground on replacing the shuttle with the new Orion and Ares I systems. Even so, Orion will not be operational until FY14, the last year allowed by presidential policy guidance. It had been hoped by many in NASA, the White House and Congress that we could deploy Orion as early as 2012. The later delivery of this key first element in the exploration architecture was accepted precisely because no one wanted to cut the science budget in order to deploy a shuttle replacement vehicle earlier. This was a serious and significant commitment to science at NASA, one which was made in the face of very tough issues in the human spaceflight program. That commitment implies that the United States, in the face of growing international competition, will not have a human spaceflight capability of its own for at least 4 years. This was an enormous step and raises national issues far beyond any in science.

The above decisions are consistent with a long period of support for and growth in the portfolio of SMD. Science today comprises a larger piece of the NASA portfolio than ever before: 32 percent today as compared with 24 percent back in the mid-90s.

While we will still launch a mission to Mars at every orbital opportunity, we have rebalanced what many viewed as an excessive increase—about 40 percent—to robotic Mars exploration at the expense of other areas in science. Further, we have restored some cuts made previously in Earth science and sponsored a National Academy study to produce the equivalent of a "decadal survey" in this field for the first time. These decisions reflect a commitment by NASA to long-term balance in our science portfolio and recognition of the key role of Earth science in that portfolio. Earth science at NASA receives $1.5 billion annually, more than 25 percent of our science portfolio.

There has been a strong, visible and clear intent by NASA management to restore the previously cancelled Hubble servicing mission, if it is technically possible to do so. A final decision and an accompanying announcement should be made by November.

In support of both National Academy priorities and long-standing international commitments, we have the reviewed the Stratospheric Observatory for Infrared Astronomy (SOFIA) mission, restored funds to the program and redirected the management strategy so as to offer the greatest possibility for ultimate success, despite a history of significant overruns and schedule slips.

We've completed the Earth Observing System with the recent launches of Cloudsat and the Cloud-Aerosol Lidar and Infrared Satellite Observation (CALIPSO) mission, and will be taking part in the multi-agency National Polar-orbiting Operational Environmental Satellite System (NPOESS) effort through our development of the NPOESS Preparatory Program (NPP). And we've recently placed the system integration responsibility for the Landsat Data Continuity Mission (LDCM) here at Goddard.

There's more, indeed much more, but my point is, I think, clear. These are not the actions of a science-hostile NASA, Office of Management and Budget (OMB) or President. Quite the contrary.

So, what's all the tumult and shouting about? A few key things come to mind, concerning which you probably won't be surprised: I think it comes down to money, respect and power. So let's take these issues on.

First and most obviously, despite all the good things above, "Science" was in earlier years promised more than it is actually getting. I believe that what SMD is getting is pretty good, but it isn't what was promised in the FY05 budget, as that budget was unveiled in February 2004. That is a plain fact.

The other plain fact is that no one else at NASA is getting what they were promised either! NASA as a whole will receive fully $3 billion less than planned in the 5-year runout in the FY05 budget request. But there were several "disconnects" in that plan. The shuttle and space station were under-funded by almost $6 billion. Cross-agency support programs were significantly under-funded and NASA was subjected to a government-wide 1 percent rescission of $350 million for FY06–07. We looked for savings where we could find them, but in the end it was necessary to reduce SMD by $3.1 billion and Exploration Systems by $1.6 billion to close the FY07 budget request.

At this point, let me add a necessary footnote to the above discussion. By "under-funded," I mean that during preparation of the FY05 budget it was assumed that, since the shuttle was retiring in FY10, the program would require less money for fiscal years 2008, 2009 and 2010 than would otherwise have been the case. While strictly speaking this is true, it is not nearly as much true as had been hoped! If we're going to fly the shuttle at all, it turns out that we actually still need most of the program to be there for the last flight. And, of course, we've had to take $2.7 billion in shuttle return-to-flight costs out of what little "hide" remains in the Space Operations Mission Directorate (SOMD). So, the shuttle/space station program has been reduced to 17 flights from 28. There will be little actual "utilization" of the space station for the next several years, in contrast to the original plan; we will be doing mostly "assembly." Now, of course, the science community would by and large just as soon see the shuttle/space station program cancelled outright. But at the highest levels of national government, that simply was not the decision that was made! So, logically, it is time to move on. But what the scientific community sees in all of this is a broken budgetary promise, pure and simple.

In this context, I have on many occasions heard the accusation that NASA has betrayed the scientific community because, it is said, the Vision for Space Exploration was "sold" as being "affordable," to be "go as you can pay." To many scientists, that means very explicitly that exploration is to be funded after, and only after, all prior science commitments were satisfied. The idea seems to be that, after we've done the James Webb Space Telescope, Europa, SIM, Terrestrial Planet Finder and every other mission in the pre-VSE NASA budget, then and only then can we embark upon renewed human exploration of deep space. Well, that is simply not how it works. "Affordable" does not mean that all of science is of higher priority than anything in exploration. The programs above were approved in an earlier time, with different budget assumptions for NASA. There have been very significant budget cuts and many unplanned requirements for funding since the Vision for Space Exploration was announced. The impact of those cuts cannot fall to any single entity in NASA's portfolio. "Go as you can pay" applies to all of NASA, not just to isolated pieces of its portfolio.

That's the "money" part of it. I've outlined the arguments not because I expect to obtain agreement—far from it—but because I think it's useful to get the nature of the issue frankly into the open. Science did not get, and will not get, as much as was promised only a couple of years ago, or will anyone else at NASA or will many other areas of discretionary government spending.

So now lets move on to "respect." Once the Vision for Space Exploration was announced, the science community immediately said, as if with once voice, "Robotic science is exploration too" and "Besides, exploration without science is tourism! No more flags and footprints!" (Which is to me, by the way, a rank mischaracterization of Apollo, but I won't fight that battle here. I will note that approximately one-fourth of Apollo funding was devoted to the last six scientific exploration missions to the moon, missions that resulted in a profound increase in our understanding of the history of terrestrial planets, particularly Earth, and of the environment in which it and life evolved.) I'm sure you've heard all of this and more. Since the science community had never previously characterized their work in terms of "exploration," many observers concluded that the theme underlying these view was, more cynically, "Don't cut our budget to pay for human spaceflight!"

Now, certainly exploration includes and enables science, for it opens and offers new capabilities to do exciting new science in new ways from new places, and about those new places. What an incredible opportunity!

But, as always, there is another view, best and most tersely captured by the President's Science Advisor Jack Marburger in his March 2006 speech at the American Astronautical Society's Goddard Symposium. Jack noted that the Vision for Space Exploration is fundamentally about bringing the resources of the solar system within the economic sphere of mankind. It is not fundamentally about scientific discovery. To me, Marburger's statement is precisely right.

So a key point must be made: Exploration without science is not "tourism." It is far more than that. It is about the expansion of human activity out beyond Earth. Exactly this point was very recently noted and endorsed by no less than Stephen Hawking, a pure scientist if ever there was one. Hawking joins those, including the chairman of the NASA Advisory Council, who have long pointed out this basic truth: The history of life on Earth is the history of extinction events, and human expansion into the solar system is, in the end, fundamentally about the survival of the species. So to me exploration is, in and of itself, equally as noble a human endeavor as is scientific discovery.

Now, portions of the broader scientific community feel deeply disrespected—I can think of no other word—when I, or anyone, says or implies that "Exploration" is not primarily about "Science." There exists a view that the only reason we go into space is to pursue scientific discovery. To me, that is a reason, but it is certainly not the reason.

Scientists frequently tell me that they want to "be a part" of the Vision for Space Exploration. And that is essential. But to be a part of the Vision does not mean to collect money that would otherwise go into manned spaceflight. It means rethinking planned programs of scientific activity in light of the opportunities to be made available through a newly vigorous program of human exploration. That is exactly what our NASA Advisory Council is asking the community to do with its planned Lunar Science Workshop next year.

I have said on numerous occasions—many of you have probably heard me say it—that the Vision for Space Exploration is not about getting more money for manned spaceflight. It is obvious that such is not going to occur. Rather, the Vision is about redirecting the money that the nation has been spending on human spaceflight to better purposes than we have been spending it. That is the key.

Similarly, participation by "Science" in space exploration cannot be about the transfer of money into SMD. It can only be about redirecting the money being spent in existing scientific arenas, along lines which the scientific community believes to be more productive, given the fact of human exploration and utilization of the moon, Mars and near-Earth asteroids in the coming decades. It is about refocusing our thoughts as to the merits and nature of future programs, given that humans will be operating in space beyond low Earth orbit.

This is the attitude that must prevail if there is to be respect by non-scientists for the contributions "Science" can make to exploration. And it is the attitude that must prevail if scientists are to show appropriate respect for those whose primary focus is to expand the scope of the stage upon which we humans act. If mutual respect can be developed between these two groups, they can be allies rather than adversaries in the grandest endeavor I can imagine. Scientists and non-scientists alike must remember that "exploration science" is not an oxymoron.

Finally, there is the issue of control. Many members of the scientific community fully understand that the president and Congress have made decisions about the shuttle and space station programs that will not be undone. They understand that the proportion of funding at NASA that goes to SMD is at an historic high, and that they should pocket their gains over the last decade and remain quiet, lest someone notice! They understand that NASA is unlikely to grow in real terms, and that therefore many projects which all of us would like to do earlier will, in fact, be done later. They get all of that.

The problem is that these folks do understand these real-world limitations, and in a world with such limitations, they want to be in charge of the distribution of resources. Put bluntly, they want to exercise the inherent authority of government to decide what is being done with the money which is available for science at NASA. But, they don't want to come to Washington, put on a NASA badge, make all the associated sacrifices and live with the consequences of their decisions, which mostly means that when you decide to do one thing, you are also deciding not to do something else that someone else would like to do, and you have to be publicly accountable for that fact.

This is the world of the many advisory committees and groups that rendered guidance to NASA, especially to NASA/SMD before I became administrator. Some of these external folks really seem to believe that NASA program selection and planning should be vetted through "the community" for approval. It is one thing to say that, broadly, we should be guided by the decadal plans of the NAS, the organization to which Congress looks for strategic advice in such matters. I emphatically support this view, while also being of the belief that sometimes circumstances change on time scales shorter than a decade, and also that sometimes good advice comes from other directions. But it is another thing entirely to suggest that "the community" has an inherent right to review and modify our annual budget. To me, one of the most disturbing aspects of this practice is that the very same people who stand to benefit from particular distributions of NASA funding would be advising NASA as to what those distributions ought to be.

Let us for a moment consider the situation in the abstract. The market for scientific goods and services, while dominated in the space sciences by the government, is nonetheless a market like any other. So, each year the president and Congress (mostly upon the advice of scientists) determine that the pursuit of certain goals in space and Earth science is in the best interests of the United States. Each year, Congress approves the purchase, through NASA, of scientific goods and services to that end. As with most markets, there are more parties desiring to provide such products than can be procured, and so a variety of closely supervised competitive procurement mechanisms are employed to determine the successful suppliers of these products. Thus, from a legal, contractual and managerial perspective, members of the external scientific community are suppliers to NASA, not customers.

My point is that if we were to substitute above any other noun besides "science," the inherent conflict between the role of the scientific community as a purveyor of products to the government, and its role as the primary source of advice as to which products the government should purchase, would not be tolerated. Yet, the scientific community simply must be involved if we are to set intelligent priorities among the nation's various scientific goals. The whole process is ethically defensible if, and only if, a proper "arm's length" separation is maintained between advisors and implementers.

This is a very fundamental issue, a matter of organizational governance. So where, exactly, do external advisors fit into the development and execution of the NASA science strategy? Let's review the bidding concerning NASA's advisory committee structure.

Leaving aside temporary groups established for specific purposes, such as the committees that investigated the *Challenger* and *Columbia* accidents, legislation governing NASA includes three specific groups chartered by

Congress to advise the administrator. These are the NASA Advisory Council (NAC), the Aerospace Safety Advisory Panel (ASAP) and the new International Space Station Independent Safety Task Force (ISS-ISTF). These groups examine our programs from the various perspectives suggested by their names and make recommendations to the administrator. Pursuant to the 2005 NASA Authorization Act, any recommendations from the latter two groups are also provided to Congress.

I take these advisory groups, and the importance of their roles, very seriously indeed. Recommendations from our congressionally chartered advisory groups are, and must be, considered and evaluated thoughtfully, and we at NASA must respond to them in a timely and substantive fashion, whether we choose to adopt a given idea or not. In fact, I have sought to elevate the role of the NAC relative to that it has occupied in recent years, because of its statutory role as the primary external advisory group for NASA in its implementation of national space policy.

But that's all there are. There are no other standing committees or interested parties required or permitted to review and advise NASA, no other group whose recommendations should be thoughtfully evaluated and to which the agency must respond. Now, all of you know that there are many, many individuals and groups whose interests are affected by NASA programs and decisions, and who believe that they deserve "a seat at the table" in helping to shape such decisions. But there is no foundation for such a belief.

Several independent groups and committees had been chartered by NASA, particularly in connection with the science advisory structure for SMD, prior to my tenure as administrator. They are gone. Instead, a quite similar advisory structure now exists as a group of subcommittees under the aegis of the NAC. I have done this for three reasons. First, in a "strict constructionist" sense, I prefer to use the advisory structure provided by Congress to help manage NASA. Second, mutually independent committees advising particular elements of NASA from a particular perspective can easily—I would say inevitably—offer conflicting and uncoordinated advice lacking concern for the larger perspective, with no need to resolve inherent conflicts with other portions of the agency's portfolio or direction. Third, it was my observation that NASA managers have sometimes used these advisory committees to assist in shaping the direction of our programs to a degree that I find unseemly, in view of the inherent potential for conflicts that I have outlined above, and in a manner tending to reduce responsibility and accountability on the part of NASA officials.

Bringing the more specialized advisory groups together as subcommittees under the purview of the NAC, which reports to the administrator rather than to individual organizational elements within SMD, addresses and resolves these issues. This structure offers and allows frequent interchange between NASA SMD staff and the external scientific advisory community without diffusing the responsibility of NASA managers for their programs. It also allows the NAC to weigh the advice of its Science Committee, or any other of its committees, against the perspectives and responsibilities of other mission directorates and other managerial units of NASA before making final recommendations to me. I believe this to be the proper way to provide an open forum for the full spectrum of advice and perspective that might be of utility to NASA, while at the same time allowing the NAC leadership to winnow and focus such advice in a manner deemed appropriate by the Council. And since the Council reports to the administrator, formal advice to NASA follows the formal chain of command used to manage the agency.

It has been said that in restructuring the scientific advisory committees as I have done, I have somehow "diluted" (that word from a September 2, 2006 *New York Times* editorial) the voice of the scientific community, or have otherwise attempted to stifle debate and discussion or am trying to suppress advice that I do not wish to hear.

This is nonsense. It is simply a fact that the NAC subcommittee meetings are open to, and heavily attended by, NASA managers and key staff. Anyone who attends is instantly privy to all advice and discussion that is aired, or which is working its way through the system. An obviously good idea can be adopted by a NASA manager without waiting for a formal recommendation. The public—including the media—is present for final Council deliberation and action. There is no "dilution" of advice whatsoever. There is only the question of whether a committee accepts a given piece of advice, whether the Council as a whole agrees with the committee's recommendation or whether it suggests alternative wording. Generally, the Council has gone along with committee recommendations; the one major exception occurred when the Science Committee recommended that it be able to bypass the Council, the NAC Chair and the administrator and provide "tactical" advice directly to SMD.

Let's consider this particular recommendation. How many of you present here today, and who are organizational managers at any level, would appreciate external advisors—or even other managers—bypassing you to provide "tactical" advice to those who report to you? Any takers for this approach to organizational governance? And if not, would it make a difference if the staff members and the advisors are "scientists" as opposed to other employees?

Moving on, it has also been alleged that, in reshaping the advisory committee reporting structure, I am "preventing scientists from talking to scientists." This is also nonsense. As far as I am concerned, anyone can talk to anyone and probably should! I desperately hope that the staff of NASA's SMD converses widely and frequently within the community. The NASA scientific staff absolutely must be of the scientific community, and active in it, to be effective in the planning and execution of their work. But the rendering of formal advice from an advisory committee to officials of a federal agency is hardly "scientists talking to scientists," nor should it be.

In fact, with regard to scientific advisory committee input to NASA, the real issue is not whether "scientists can talk to scientists," but whether the administrator is to be included in the conversation! By requiring formal advice to be debated in and provided through the NAC, the scientific community's advice to NASA comes to the administrator and simultaneously to SMD. Under the prior structure, with numerous committees reporting directly to lower-level organizational managers, the administrator usually had no direct knowledge as to the advice being provided to the agency by external groups. This is not a responsible approach to organizational management.

Thus, at this point, I am back to basic organizational management principles. Responsibility and accountability for planning and executing NASA's science program must rest with NASA's managers, not the external scientific community. Execution of these responsibilities must be appropriately informed, and to this end we must, and will, make intelligent use of our advisory committee structure. But the final responsibility and accountability for agency programs can lie nowhere other than with us, the NASA staff.

I hope I have been able to clarify my thinking with respect to how science fits into NASA's overall strategy. I am deeply committed to having a robust science portfolio, and my actions have been consistent with that commitment.

As administrator, I put the Hubble Servicing Mission back into our science plan. I rebalanced the science portfolio out of respect for National Academy priorities and out of concern for the health of important disciplines like Earth science and heliophysics. I did have to cut the growth rate for science, but other parts of the NASA portfolio, including exploration and aeronautics, have made similar sacrifices.

Others with more singular and self-interested views of NASA's purpose would like to divide and conquer us. They would like to cast the argument in the terms "Science verses Exploration." That argument is deliberate and deceptive. I don't accept it, and I urge you to reject it as well. The Vision for Space Exploration was wise to call for the use of both robotic scientific missions and human scientific missions in the exploration of the universe. It rightly recognized the strengths of both endeavors, and it understood the symbiosis between robotic science and human exploration that will characterize our exploration campaigns.

So, this isn't about "Science versus Exploration." We will do both. And we will succeed with both. Both will contribute greatly to increased understanding of ourselves, the environment in which we live and the solar system and universe around us. And because of the mutually reinforcing relationship between the two, we will do both better than we could do either one alone. This will be a productive partnership, and the sooner we recognize that, the better that partnership will be.

And finally, as I have said from day one, we (NASA) are responsible for executing the nation's space program. Sixteen months ago, shortly after joining NASA, I was asked if I would approve the flight of STS-114 in the face of concerns by some members of the Stafford-Covey Return to Flight Task Group. I said then that the role of advisors is to advise; NASA decides. I will no more submit NASA's scientific decision making to external committees than I submitted NASA's shuttle flight readiness decisions to them. And I say that with no disrespect to NASA's important advisors. When they have something to say, they should say it; and we should listen and listen carefully. But in the end NASA (and by "NASA" I mean you and I) is responsible for the decisions of this agency. And you should understand that in taking that position, I am not only committing to a certain kind of governance in which I strongly believe, but I am also demonstrating commitment to and respect for you.

Thank you for choosing to spend your time with me today.

Science Priorities and Program Management

Michael D. Griffin
Administrator
National Aeronautics and Space Administration

Remarks to the NASA Advisory Council
Science Subcommittees

July 6, 2006

We celebrated our nation's 230th birthday, our Independence Day, this week. One of the great strengths of our country is the principle of freedom of speech, of entertaining vigorous debates on the great issues of the day. For NASA, the great issue before us is how we carry out our nation's civil space program—in space science and human spaceflight—and our aeronautics research programs.

We are a nation of laws, and to that end NASA is governed by the Space Act of 1958 as our founding charter, just as the Declaration of Independence and the Constitution are the founding charters of our nation. The NASA Authorization Act of 2005 and presidential policy—the Vision for Space Exploration—provide the long-term direction for our investments of time, resources and energy in the nation's space and aeronautics program. Each year's budget and appropriations legislation provides detailed guidance in crafting an overall portfolio of missions in space exploration, scientific discovery and aeronautics research. Thus, all NASA programs are "go-as-far-as-we-can-afford-to-pay" at the national level.

When it comes to space science priorities, we are guided by the decadal surveys of the National Academy of Sciences, and I'm glad that we'll soon have a decadal survey for Earth science priorities. This brings us to your role. We've specifically asked for your collective and personal advice as to how we carry out NASA's science programs—astrophysics, heliophysics, planetary science and Earth science. In this town, advice is often freely given, but in your case, we're actively seeking it. You are some of the most senior representatives and emerging leaders of your respective fields of endeavor. So, I offer my thanks to you all for agreeing to be part of the NASA Advisory Council's science subcommittees.

One of the issues where we need your advice concerns the fact that human exploration of the moon, Mars, near-Earth asteroids and the rest of the solar system is not solely science-driven. However, given that this effort will be undertaken, we are seeking the counsel of the science community as to what science can be done in the course of the human exploration of the solar system. Jack Marburger framed this issue very well in his speech in March at the 44th annual Goddard Memorial Symposium. "The question about the Vision boils down to whether we want to incorporate the solar system in our economic sphere, or not."

Our national policy, declared by President Bush and endorsed by Congress last December in the NASA Authorization Act, affirmatively answers that question: "The fundamental goal of this vision is to advance U.S. scientific, security and economic interests through a robust space exploration program." Scientific discovery through human exploration is one goal of the Vision for Space Exploration, but is not the only goal.

We will definitely add to the scientific body of knowledge for our civilization about the real estate values in the vicinity of Earth, and we will conduct scientific experiments along the way, much in the fashion that Meriwether Lewis and William Clark gathered specimens, made careful observations in their journals, and drew detailed maps of the great American West 200 years ago.

Like Lewis and Clark's maps of the newly-acquired territories that expanded our nation's economic sphere, NASA's Lunar Reconnaissance Orbiter will provide detailed terrain elevation data for future exploration and use of the resources of the moon, just as the Mars Global Surveyor and Mars Reconnaissance Orbiter are mapping details of the surface of that planet in the search for evidence of potential subterranean flowing water and future landing sites for robotic and human exploration. But what other scientific endeavors should we pursue in low-Earth orbit, or on the moon, Mars and near-Earth asteroids during the course of human exploration of our solar system over the next several decades?

We need the scientific community to help us answer that question.

As we organize our endeavors, I'd like to call your attention to the recommendations from the 1990 decadal survey in astronomy and astrophysics of the National Academy of Science, commonly known as the Bahcall report. Back in 1990, this National Research Council (NRC) committee studied the suitability of the moon for possible astronomical facilities and found that, in the long term (though not even in the next decade), the chief advantage of the moon as a site for space astronomy was that it provided a large, solid foundation on which to build widely separated structures such as interferometers.

This same report from 1990, along with the next decadal survey, "Astronomy and Astrophysics in the New Millennium" in 2000 and the annual report from the NRC's Astronomy and Astrophysics Advisory Committee have helped me to make some difficult decisions recently by informing me as to how the science community viewed certain priorities. In 1990, both SOFIA and Astrometric Interferometry Mission, since renamed the Space Interferometry Mission (or SIM), were regarded as moderately priced programs compared to large programs like Space Infrared Telescope Facility (SIRTF)/Spitzer. Obviously, this survey underestimated the complexity, cost and schedule for both of these projects; but the decadal survey ranked the Stratospheric Observatory for Infrared Astronomy (SOFIA) as a higher science priority than SIM.

Even before I became NASA administrator in the spring of 2005, I knew of problems with the SOFIA program, due to gross underestimates of the technical complexity of integrating a 2.5-meter telescope onto a Boeing 747 airborne platform. Costs grew to the point of making SOFIA a large program and the schedule kept slipping further to the right.

Earlier this year, I believed that the best course of action at that time was to withhold funding in fiscal year (FY) 2007 for SOFIA, until we conducted a thorough review and carefully considered the next steps for this project. That review included the option of terminating it. Having received this report, I now believe the best course going forward is to continue SOFIA, with some significant management changes.

After a careful and independent technical and management review this past spring, NASA's Program Management Council concluded that the remaining technical challenges for SOFIA, like the stability of the telescope within the aircraft's cavity door drive system, were not insurmountable. However, we decided that we needed a team in place to manage SOFIA having a greater level of management experience with research aircraft. Thus, we decided that the Dryden Flight Research Center should lead the development and flight tests of SOFIA. We also need to simplify the contracts to ensure that Dryden project managers have direct authority over the contractors actually performing work on the aircraft. Dryden operates several research aircraft and has considerable hands-on experience with such issues. Ames Research Center will continue to retain science management responsibility for SOFIA, though we may later re-evaluate the science management responsibilities as the project continues.

Following the NASA Program Management Council's (PMC) technical and management review, we then sought to determine whether SOFIA represents a better investment for space science funding than other projects in the universe/astrophysics portfolio. For this analysis, we were informed by the 1990 Bahcall report, where SOFIA was ranked as a higher priority than SIM, the 2000 decadal survey which reaffirmed those recommendations and the NRC's annual report from the Astronomy and Astrophysics Committee last March, which said: "With a substantial expenditure on Hubble Space Telescope servicing, increases in James Webb Space Telescope's construction cost and significant funding for SIM (despite its not earlier than 2015–16 launch date), the astrophysics program is overly biased towards large missions. The science return from such missions is not in doubt, but the lack of balance will impact future opportunities and the diversity of scientific investigations. As discussed in more detail in the report, substantial delays in the shuttle availability for HST SM4 (Hubble Space Telescope Servicing Mission 4), any further cost growth in JWST and the funding profile for SIM are all issues that need to be considered. (SIM has a high lifecycle cost because of both current significant spending and early ramp-up.)"

In addition to SOFIA, the universe/astrophysics theme has a diverse portfolio of projects, ranging from "Great Observatory" missions like the Hubble Space Telescope, James Webb Space Telescope, SOFIA and SIM as well as smaller-class missions such as Kepler, Wide-field Infrared Survey Explorer (WISE), Gamma-ray Large Area Space Telescope (GLAST), Herschel and Planck. Despite losing the opportunity to make some observations in conjunction with the Spitzer Space Telescope, SOFIA can still fulfill its science objectives to a degree commensurate with our investment and has the potential to produce "Great Observatory" science over its 20-year design life. As an airborne platform more readily able to incorporate and test a wide range of astronomical instruments than a space telescope, SOFIA has a great deal of flexibility and can benefit a broad range of astronomers while complementing the capabilities of NASA's space telescopes. As a research aircraft, SOFIA can also provide hands-on training and education for future astronomers.

Thus, in order to continue SOFIA out of the $1.5 billion spent per year in NASA's astronomy and astrophysics portfolio, we plan to refocus SIM as a technology and research effort for finding Earth-like planets in other solar systems, a portfolio of projects which includes the Kepler Space Telescope. NASA will then be informed by the priorities in astronomy and astrophysics from the upcoming decadal survey, to be initiated in 2 years. I have made this consideration carefully, and I believe that it is the best course of action for SOFIA as well as the rest of the astronomy and astrophysics portfolio.

As I have told Charles Elachi, I am sensitive to the impacts to scientists and engineers at the Jet Propulsion Laboratory (JPL) who worked on SIM. I have been laid off twice in my career. He has my commitment to maintain JPL's overall workforce at its present level, with the assignment of other work as necessary to do so. However, the priorities of the decadal surveys in astronomy and astrophysics are clear, as is the advice of the NRC's annual report. Clearly, we in the broader space community have a credibility problem with our stakeholders in managing the technical complexity, costs and schedule for our programs. The science community must be careful not to underestimate the costs and complexity of the missions it proposes. NASA cannot afford everything our many constituencies would like us to do.

However, having been a part of several Nunn-McCurdy breach reviews for NPOESS and reading reports concerning other major DOD programs, this problem is broader than NASA, and I have discussed it extensively with former Congressman Dave McCurdy, Tom Young, my counterparts in the DOD and members of Congress. Our stakeholders in Congress are concerned that NASA not under-estimate the costs or complexity of our programs. To that point, the NASA Authorization Act of 2005 requires even more stringent management actions than those in the Nunn-McCurdy legislation for NASA missions costing more than $250 million and which exceed their base-lined costs. I would ask everyone in the science community who proposes missions to NASA to become familiar with that legislative provision, which is now the law of the land and which I and my managers must follow. In the future, decisions such as whether or not to continue missions like SOFIA will not be left to the NASA administrator but will go to Congress.

My number one request to Mary Cleave Associate Administrator for the Science Mission Directorate and Colleen Hartman Deputy Associate Administrator for the Science Mission Directorate is that they bring forward realistic, executable programs within their budget. We will need your help in this. Every time we have over-promised on a program, we have lost credibility with our stakeholders.

Speaking of stakeholders, it's now the time of the year when the House and Senate committees mark-up their bills for the following year's appropriation. Recently, the chairman of NASA's appropriation subcommittee, Congressman Frank Wolf, displayed real leadership by curtailing individual member's earmarks for NASA and the National Science Foundation (NSF). "One person's priority is another's earmark," Jack Marburger points out, and he's absolutely right. I believe the science community is best governed by merit-based, peer review procedures. Our hope is that the science community can form a consensus on its priorities, as with the decadal surveys, which would argue against funds being diverted for one person's earmarks. Back in FY97, specific direction for NASA constituted only $74 million for six specific projects. In FY06, NASA was earmarked at a total of $568.5 million for 198 site-specific and programmatic increases, with $48.3 million in site-specific earmarks from NASA's Science Mission Directorate and $63.4 million from our Education programs.

As members of the science community, we need your help in curtailing this level of earmarking among your colleagues. NASA simply cannot afford everything that everyone would like us to do. Chairman Wolf recognizes these difficult choices and the need to focus limited resources on programs most critical to our nation. We are working closely together on this. I have also discussed with Chairman Wolf the need for more discussions within the planetary science community to set priorities for missions to the outer planets and moons of Jupiter and Saturn. These missions will cost a minimum of several billions of dollars. While a mission to Europa was ranked as the highest planetary science priority in the decadal survey published in 2003, since then we have learned that liquid water might also be found on Enceladus, one of Saturn's moons; and Titan also has an interesting methane-rich atmosphere with volcanic activity. Neither of these two moons has a harsh radiation environment like that of Europa, whose extreme radiation field could cripple a multi-billion spacecraft in its orbit before it completed its science mission. Thus, I believe that the best course of action moving forward is to permit the science community to determine the next outer planets mission through a competitive selection process under the New Frontiers program.

I would also like to note for the science community that, if you advocate large missions exceeding the capabilities of the current Evolved Expendable Launch Vehicle (EELV) fleet, you should consider taking advantage of new heavy-lift capabilities currently under development for human exploration. While the planetary science community may not have liked my decision, as part of the FY07 budget formulation, to place the national priorities of completing the space station, retiring the space shuttle by 2010 and bringing the CEV on line no later than 2014, higher than the goals of missions to the outer planets like Europa, I want to assure you that our nation will carry out such missions. It simply will not occur as soon as some might wish that mission to be. Does that make me less of a fan of missions to the outer planets? Absolutely not. I'm trying to put forward an affordable and credible portfolio of missions within NASA in accordance with the law of the land and national policy and avoid making promises the agency cannot keep. I strongly believe this to be in the best interests of the overall space program.

These are the issues at the forefront of my mind today as I look out at the landscape of NASA's broad portfolio of science missions. We're asking for your advice on the journey ahead. This has been a momentous week for NASA, with the second Return-To-Flight shuttle mission underway on July 4; and I'm glad to be spending some time with you now. After this mission is completed, I will convene a group of senior NASA managers to help me decide the best course ahead for a servicing mission with the shuttle to the Hubble Space Telescope.

Meriwether Lewis observed the following perspective in his journal on July 4, 1805 which speaks across two centuries to many of us in NASA: “We all believe that we are now about to enter on the most perilous and difficult part of our voyage, yet I see no one repining; all appear ready to meet those difficulties which wait us with resolution and becoming fortitude.”

We have a lot of work to do. We are asking for your advice as to how we carryout that work. Let me now open this dialogue to your questions.

NASA and the Commercial Space Industry

Michael D. Griffin
Administrator
National Aeronautics and Space Administration
Remarks at the X-Prize Cup Summit
October 19, 2006

An insightful, and often all too apt, observation goes: "There are three types of people in the world: people who make things happen, people who watch things happen and people who wonder what happened." The group assembled here clearly fits into the first category, and so for my part, I'd like to spend some time with you this morning wondering what happened. More seriously, I believe this observation needs a fourth category ahead of the three given above; first there must be the people who think about what ought to happen. These are the visionaries, and none of us would be here at this event today without them. So, I want to spend some time with you thinking about what needs to happen next.

All of you here will be familiar with our new Commercial Orbital Transportation Services (COTS) demonstrations being conducted under the framework of NASA Space Act Agreements. These landmark agreements are, truly, NASA's most significant investment to date in attempting to spur the development of the commercial space industry. But let me say this at the outset: NASA can do even better in partnering with the commercial and entrepreneurial space sector in carrying out our nation's Vision for Space Exploration. However, let me be equally blunt about the other side of the coin: "partnership" with NASA is not a synonym for "helping NASA spend its money." Just as with our international partnerships, I expect commercial and venture capital partners to have "skin in the game," contributing resources toward a common goal that is greater than that which could be easily afforded by NASA alone, while figuring out how to make a profit from it!

Thus, it is important for the future that NASA's investments productively leverage the engine of the American economy, a Gross Domestic Product (GDP) valued at over $13 trillion per year, to help us carry out our mission of space exploration. As the President's Science Advisor Jack Marburger stated earlier this year, "questions about the Vision boil down to whether we want to incorporate the solar system in our economic sphere, or not." I think that I can guess how most of you who are here today would answer that question. And, indeed, I have said in other venues that for me also, this is one of the core principles justifying human exploration and expansion into space.

But the kind of things we need to do have been done before. We know how it should go. Many of you have in the past heard me allude briefly to the story of how the U.S. Post Office Department, with the help of the War Department, helped spur our nation's aviation industry when it started the airmail service routes in 1918. I very strongly believe that we can, and should, draw certain lessons from this event; that it can be an historical paradigm for how NASA might fill a similar role in spurring our emerging commercial space industry in concert with the goals of the Vision for Space Exploration. However, a review of this history shows that it was not an easy proposition then, and it is likely to be just as difficult to pursue in the present era. But, as President John Kennedy said at Rice University in 1962, we do these things, "not because they are easy, but because they are hard." So let us look again at what was once done, and then let us think about what might yet be done.

The idea of an airmail service in the United States was initially proposed by the Post Office Department in 1912. However, Congress refused to grant them the $50,000 appropriation needed to start. Undaunted and persistent, the Post Office Department kept requesting funds from Congress for an airmail service. Finally, in 1916, some funding was received, but when the Post Office Department invited bids for airmail routes in Massachusetts and Alaska, no company took them up on their offer, because no airplanes then in existence could meet the stringent requirements. Revising its plans, the Post Office Department and the Army finally demonstrated the first airmail route between Long Island, New York and Washington, D.C. in May 1918. It was a momentous occasion, and President Woodrow Wilson greeted the pilot upon landing. Today, if you walk along the Potomac River not far from the Jefferson Memorial, you will find several plaques commemorating those first airmail flights.

Using initially the then-plentiful Curtis Jenny trainers, surplus from the Great War, transcontinental airmail routes were quickly established. By the mid-1920s, the Post Office Department's fleet was flying 2.5 million miles annually, delivering 14 million letters. This airmail service was popular because delivery times were much faster than could be accomplished using trains. However, there were also many fatalities during this barnstorming era. Cross-country flights in all kinds of weather and lighting conditions presented new and unsolved problems. The reason why pilots wore goggles and scarves in those open-air cockpits was hardly to look dashing. The goggles prevented bugs from striking the pilot's eyes at 100 miles per hour, and the scarf was to cover the pilot's mouth from the bugs that might fly in and to wipe away oil sputtering from the plane's engine. Those of us who flew here to Las Cruces today should not take for granted our current level of aviation safety and comfort. Today, you have about the same chance of being killed by a lightning strike—about 100 Americans per year die this way—as in an air transport accident. But back then, aviation accidents and deaths were all too common.

In 1925, the Contract Air Mail Act (or Kelly Act) authorized the postmaster general to contract for airmail services, and in the process spawned our nation's nascent airline industry, as the airlines delivered both paying passengers and cargo. Charles Lindbergh was one of those early pilots, flying the route between Chicago and St. Louis in his de Havilland DH-4. His experiences flying the mail in these early years—including the bailouts and emergency landings—are recounted with both great literary grace and a pilot's sense of immediacy in *We,* his autobiographical summary of those years. Lindbergh's early experiences flying the mail gave him the experience he would need for his famous first non-stop flight from New York to Paris in 1927, winning the $25,000 Orteig Prize for himself and his backers.

But the story doesn't end there. In 1933, President Roosevelt's postmaster general found unethical behavior in the awarding of these airmail contracts, and the president summarily canceled all such contracts and ordered the Army Air Corps to step in and take over the airmail service for a brief time. A young man by the name of James Webb, who was a lawyer and also a Marine Corps aviator himself, and who would later become NASA's greatest administrator, was closely involved in bringing order out of that chaos to restart the commercial airmail service. This phase of Webb's life is chronicled in his biography, *Powering Apollo,* by W. Henry Lambright.

So what are the lessons to be gleaned? First, the U.S. government acted through the Post Office Department as a major purchaser of potential air transport services, as opposed to being a technology developer. The aviation industry used the government's investment to develop their commercial operations further, and along the way, incorporated numerous technical innovations that proceeded from the Ford tri-motor to the Boeing 247 and eventually to the Douglas DC-3, generally considered to be the first practical commercial transport aircraft. These investments in soliciting actual airmail service, rather than in technology development itself, spurred innovation in retractable landing gear, radio navigation aids, aluminum monocoque structural design for low weight, low drag airframes, air-cooled radial engines, vacuum gyroscopes and a slew of other technologies, while also delivering the mail which was of course the intended primary goal. Technology development was the byproduct of this investment; it occurred as a natural result of competitive entrepreneurs attempting to out-do each other in servicing a known government market.

But second, we should remember that even as the Post Office Department was stimulating the development of aviation by purchasing commercial service, another arm of the U.S. government was doing its part from a different perspective. Aviation technology development was extensively aided and abetted by the activities of the National Advisory Committee for Aviation, or NACA, the predecessor of today's NASA. Through its three research centers—first Langley, then later the Lewis and Ames laboratories—the NACA sponsored much of the groundbreaking technology development and proof-of-concept work, providing a base of feasible technical alternatives upon which industry could draw with each new airplane design. In my opinion, this private-public synergy achieved results both far better, and much faster, than either approach alone could have done.

So, what about space? We now have more than 50 years of investment, through both NASA and DOD, in space technology and systems development. But what we have not had is a stable, predictable government market for space services sufficient to stimulate the development of a commercial space industry analogous to that which was seen in the growth of aviation. My hope is that with the seed money we are putting into the COTS program, we can demonstrate the possibility of commercial cargo and crew transportation to the International Space Station, and that subsequently NASA will be able to meet its space station logistics needs by purchasing these demonstrated services. If we can do this, we will be able to change the paradigm for transportation services to be more in line with the airmail service of the 1920s, meeting the logistics needs of the space station, some 7,000 to 10,000 kilograms per year, after the space shuttle is retired in 2010. In the process, we may be able to spur innovation for low-cost access to space. This is a carefully-considered investment with known risks that we can all see and appreciate, but with a potentially huge upside that makes it well worth the risks.

I'll risk repeating myself to ensure that everyone fully understands how serious NASA takes the COTS demonstrations: if these commercial service capabilities are successfully demonstrated and cost-effective, NASA will welcome and use them. That is our default strategy for space station resupply. Most of you will probably agree that meeting or beating the government's cost to provide space transportation services shouldn't be too difficult for private industry to do. I hope you are right. I want these demonstrations to succeed; however, my wanting it won't make it so. If these capabilities are not successfully demonstrated, then NASA's fallback position is to rely on the Orion Crew Exploration Vehicle or international partner cargo and crew service capabilities for space station logistics support.

Now, there is another lesson to be derived from the airmail experience. For the space transportation services we seek, certain human rating and visiting vehicle requirements applicable to the space station must be respected. To that end, we're interested in hearing from potential commercial providers, like SpaceX and Rocketplane Kistler, as well as Lockheed Martin's Orion team, concerning what requirements are necessary and value-added, and which ones may not be. The definition of human-rating is not simply how much paper and process we can afford to buy. That is the wrong metric. For this reason, we are reviewing the visiting vehicle and human rating requirements, not only for the COTS demonstration but also for the Crew Exploration Vehicle, to ensure that we're writing our engineering specifications to achieve the goal of technical excellence and are not simply following a handbook. Good engineers do not simply quote requirements from handbooks; we understand the underlying technical necessity behind such requirements.

Similarly, we must avoid relying solely on precedent, upon the mentality of "that's the way we did it on shuttle," or space station, or Apollo, or *Skylab* or whatever as a substitute for good judgment. If we don't periodically question our technical requirements, if we focus on process to the exclusion of outcome, if we substitute methodology for intent, then we will replicate the experience of the Post Office Department in its initial request for bids on airmail service: commercial industry will never be able to meet NASA's stated needs. Thus, we must focus upon, and be experts in, systems engineering as we work through various technical issues for our future crew and cargo systems. We must be prepared to question our assumptions when necessary.

Yet another lesson gained from the airmail service was how it helped train a new generation of pilots like Wiley Post and Charles Lindbergh, engineers like Glenn Curtiss and Donald Douglas and lawyers like future NASA Administrator Jim Webb. This barnstorming era engendered a certain sense of "air-mindedness" among the American people in much the same way that space tourism is rekindling an interest in space travel for the American public, over and above that which NASA accomplishes today. Of course, the physics and engineering are more difficult for personal space travel than for air travel, with even greater levels of cost and risk, but we must recognize that this change is occurring. There are now emerging certain rudimentary commercial capabilities for members of the public to have their own personal "space experience," with varying degrees of weightlessness and views of Earth and space. I fervently hope that the emergence of such capabilities will help make America more "space minded."

Now, I must be clear that the development of space tourism is not a part of NASA's charter. NASA was founded during the Cold War, soon after the launch of Sputnik, when the United States was in a race with the Soviets. NASA and the early civil space program were instruments of American preeminence in the world at a time when an important component of such was seen to be preeminence in space. NASA achieved the goals that were set for it by the nation's policymakers in that era, and did so with remarkable brilliance, so much so that even today we remain in awe of what the Apollo generation did. Now, some have since posited that NASA somehow failed the American public by not opening up the experience of space travel to the broader population.

This is patent nonsense; the agency could not fail at something it was never asked to do. Such a mandate was simply never in NASA's charter; if it were, I would question the wisdom of such a role for a government entity. However, as we go forward with the Vision for Space Exploration, it emphatically is our duty to encourage and leverage nascent commercial space capabilities. Not only is it the right thing to do in a country whose economic system is rooted in free market concepts, but it will also be a necessity if we are to achieve the goals set out for the U.S. civil space program.

A little over a year ago, I unveiled to Congress and the public NASA's architecture for returning to the moon. It is a conservative plan designed to accomplish the stated mission with minimum cost, maximum cost confidence and as much use of existing systems as we could reasonably achieve. But having combed through the design trades, associated costs and projected budget for the agency, it is apparent that NASA will need to leverage commercial and international partners to the maximum if we are to sustain this long journey, with footholds first on the space station, then on the moon and from there onward to Mars. It is out of necessity for, not charity toward, commercial space endeavors that we at NASA must change our way of doing business. While I think that the $500 million we're investing in the COTS demonstrations is a sizable first step, there's more gold to be mined in other fields of commercial endeavor as well.

To that end, we are taking a hard look at our government-operated microgravity research aircraft at Johnson Space Center and at what NASA requirements commercial providers can meet. We've purchased some commercial research flights from Zero-G Corporation in the past and, going forward, we are looking to meet the full set of our requirements through the purchase of private sector services at a lower cost. You recently saw a NASA Request for Information on micro-gravity flight services, and you can expect to see more from NASA in the coming months.

Commercial aircraft can make parabolic flights for 20–30 seconds of weightlessness at a time. I hope that future suborbital flights will soon be taking paying passengers to the edge of space for approximately 4 minutes of weightlessness, as well as a great view of Earth from the edge of space. Using the airmail paradigm, NASA will purchase seats for these suborbital flights for certain experiments, and possibly astronaut candidate proficiency, if and when they become available. Just as NASA pilots fly T-38s and micro-gravity aircraft flights to maintain proficiency, we should consider how we might use these future suborbital flight opportunities. I have asked NASA Associate Administrator Rex Geveden to look into this capability under NASA's Innovative Partnership Program. Rex also oversees management of NASA's Centennial Challenge Prize program, authorized by Congress last December. Several NASA prize challenges, like the lunar lander, will be featured here at the X-Prize Cup over the next several days. The spirit and heritage of these prizes harks back to Charles Lindbergh's successful bid for the Orteig Prize in 1927; I hope these new prizes spark similar accomplishments.

In another vein, the NASA Authorization Act of 2005 also designates the U.S. segment of the space station a national laboratory. NASA is actively seeking commercial partners who would like to use the space station for their own experiments. After the loss of Space Shuttle *Columbia*, NASA was forced to curtail a great deal of space station research, and with our focus on the use of the space shuttle system for space station assembly over the next few years, I believe that commercial cargo and crew services will prove invaluable for increasing access to space and to the space station for these commercial experiments.

Also in connection with the space station, we need to be open to novel concepts, which can enhance the utility of this multibillion dollar facility. As one example, former astronaut and present-day entrepreneur Franklin Chang-Diaz, creator and proponent of the Vasimir electric propulsion concept, has opened discussions with NASA in connection with the possible use of the Vasimir engine for space station orbit maintenance. We don't know, yet, whether this particular approach makes sense or not, but if it does, there might be a classic "win-win" strategy here; we gain experience with a potentially useful space propulsion concept, and we reduce the amount of propellant delivery needed for space station reboost, leaving room in the logistics manifest for more productive cargo. This is the kind of private-public synergy that can serve us well.

While we are on the theme of innovative approaches to commercial space endeavors, I want to congratulate Pete Worden and his team at Ames for working with Bigelow Aerospace to secure a piggyback ride for their Genebox experiment on Bigelow's Genesis inflatable space habitat demonstration. I believe that this is one of many innovative, short turnaround ideas that we'll be seeing from Pete over the next several years. He is turning Ames Research Center in Silicon Valley into a "Mecca" for space entrepreneurs, where among other things we are hosting the Red Planet venture capital fund, similar in some ways to the CIA's In-Q-Tel operation, to leverage innovators and investors who have not typically done business with NASA.

Recalling again the lessons of the airmail service in 1933, we know that we must avoid any real or perceived favoritism before entering into any joint endeavors. There must be a healthy competition of ideas and resources. Before making commitments, we must carefully consider and ensure that joint endeavors are properly aligned with NASA's mission, sufficiently high priority and can be done within the resources provided to NASA. Now, I specifically want to emphasize that the phrase "carefully considered" is not a euphemism for hiding behind bureaucratic processes or legalistic red tape. If you see this happening, we want to hear about it. Having worked in industry, I appreciate the need to meet a payroll, and I know well how the timing of government decisions affects your "skin in the game." For this reason, clear dialogue is necessary between NASA and the parties involved when exploring possible joint endeavors. We must not over-promise or over-commit. It is one thing to begin an endeavor, but it is an even greater accomplishment to complete it! Too many exciting endeavors at NASA have failed to meet this standard in recent years. We must re-establish NASA's reputation for finishing what we start.

As I stated earlier, there are people who make things happen, people who watch things happen and people who wonder what happened. I'll share with you another of my favorite aphorisms: managers do things right, but leaders do the right things. So, we need to make things happen, but we also need to make sure that we're trying to make the right things happen. The lessons learned from our nation's first steps in creating a commercial airmail service are useful to us today. So, let me leave you with a final thought from a certain airmail pilot, one Charles Lindbergh: "It is the greatest shot of adrenaline to be doing what you have wanted to do so badly. You almost feel like you could fly without the plane." The group assembled here today knows that feeling. So, let's make things happen, so that we can enjoy it more often!

NASA and the Business of Space

Michael D. Griffin
Administrator
National Aeronautics and Space Administration
American Astronautical Society
52nd Annual Conference

November 15, 2005

When President Bush announced the Vision for Space Exploration in January 2004, he made many specific points, including one which has been little noted but which we here all believe: the pursuit of the Vision will enhance America's economic, scientific and security interests. He also made it clear that the first step in the plan was to use the space shuttle to complete the assembly of the International Space Station, after which the space station would be used to further the goals of exploration beyond low Earth orbit. These issues are all closely related, and I believe it is time to discuss in more detail how the space station will be used to accomplish them, and how it will fit into a broader strategy for 21st century space exploration of the moon, Mars and beyond in a way that will spur commerce, advance scientific knowledge and expand humanity's horizons.

We are entering the dawn of the true space age. Our nation has the opportunity to lead the way. It is an opportunity we are eager to pursue, and one which we are unwilling to postpone. But the exploration of the solar system cannot be what we want it to be as an enterprise borne solely by the American taxpayer, or even by the taxpayers of the nations willing to join with us in this enterprise.

If we are to make the expansion and development of the space frontier an integral part of what it is that human societies do, then these activities must, as quickly as possible, assume an economic dimension as well. Government-directed space activity must become a lesser rather than a greater part of what humans do in space. To this end, it is up to us at NASA to use the challenge of the Vision for Space Exploration to foster the commercial opportunities that are inherent to this exciting endeavor. Our strategy to implement the Vision must, and we believe does, have the potential to open a genuine and sustainable era of space commerce. And the space station will provide the first glimpses into this new era.

Before we pursue this thought further, let us summarize a few statistics from the space station program. On November 2, 2005, we marked the fifth year of consecutive human occupancy of the station. The station has hosted 97 visitors from 10 countries in its approximately 425 cubic meters, a volume roughly the size of a typical three-bedroom home. Of these, 29 have been crewmembers of the 12 space station expeditions that have flown to date. With the most recent spacewalk by Expedition 12 Commander Bill McArthur and Flight Engineer Valery Tokarev, 63 have been conducted in support of space station assembly, totaling nearly 380 hours. And through the partnership we have with 15 other nations, we have learned to work together on an incredibly complex systems engineering project. While it certainly has not always gone smoothly, the simple fact of its accomplishment has been an amazing feat. My oft-stated view is that the international partnership is, in fact, the most important long-term benefit to be derived from the space station program. I think it is a harbinger of what we can accomplish in the future as we move forward to even more ambitious objectives in space.

Indeed, the value of this international collaboration was endorsed once again by a recent vote in Congress, which lessened certain restrictions placed on our ability to cooperate with Russia in the arena of manned spaceflight. This congressional action helps to ensure the continuous presence of American astronauts on the station. It continues to reflect our government's commitment to nonproliferation objectives, while recognizing the value of international cooperation in space exploration.

So, how can the space station that we are building today help us to move beyond low Earth orbit tomorrow?

To begin, we are focusing human research on the space station on the highest risks to crew health and other issues we will face on long exploratory missions. This research will help us understand the effects of long duration spaceflight on the human body, such as bone and muscle loss, so that we can develop medical standards and protocols to manage such risks. We have already had some successful anecdotal experience among space station crewmembers with exercise countermeasures. Perhaps space station-based research will one day help us to evaluate the efficacy of drugs to counter osteoporosis, or long-term exposure to the radiation environment or to test advanced radiation detectors. The station will help us learn to deal with crew stress on long missions, to enable them to remain emotionally healthy.

With the space station as a test bed, we can learn to develop the medical technologies, including small and reliable medical sensors and new telemedicine techniques, needed for missions far from home. A milestone in that arena was achieved a year ago when the journal *Radiology* published its first research paper submitted directly from the station—Space Station Science Officer Mike Fincke's account of the first use of ultrasound in space for a shoulder examination.

The space station can host and test developmental versions of the new liquid-oxygen (lox)/methane engines we will need for the Crew Exploration Vehicle (CEV), and many other systems that we will need for Mars. These include the development and verification of environmental control, life support and monitoring technologies, air revitalization, thermal control and multiphase flow technologies, and research into flammability and fire safety. As I have often said, when we set out for Mars, it will be like sealing a crew into a submarine and telling them not to ask for help or return to port for several years. We can't do that today. We have to be able to do it before people can go to Mars. We'll learn to do it on the space station and later on the moon. And so, fundamentally, the space station will allow us to learn to live and work in space.

And even though this research is focused on the tasks associated with setting up research bases on the moon and preparing the way for Mars exploration, it will also benefit millions of people here on Earth. What we learn about bone loss mitigation and cardiovascular deconditioning, the development of remote monitoring and medical care, and water reclamation and environmental characterization technology obviously has broader benefits. One certainly would not build a space station to achieve these goals. But given that we have it, we intend to maximize the science return from the space station in ways that will benefit both space exploration and our society at large.

But now let us turn to what I believe will be an even greater benefit of the space station, and that is its role in the development of space as an economic arena.

In order that we may devote as much of NASA's budget as possible to the cutting edge of space exploration, we must seek to reduce the cost of all things routine. Here in 2005, the definition of "routine" certainly should include robust, reliable and cost effective access to space for at least small and medium class payloads. Unfortunately, it does not, and frankly, this is not an area where it is reasonable to expect government to excel. Within the boundaries of available technology, when we want an activity to be performed reliably and efficiently, we in our society look to the competitive pressures of the free market to achieve these goals. In space, these pressures have been notably lacking, in part because the space "market" has historically been both specialized and small. There have been exceptions—notably in the communications satellite market—but the key word here is "exceptions." Broadly speaking, the market for space services has never enjoyed either the breadth or the scale of competition which has led, for example, to today's highly efficient air transportation services. Without a strong, identifiable market, the competitive environment necessary to achieve the advantages we associate with the free market simply cannot arise.

I believe that with the advent of the space station, there will exist for the first time a strong, identifiable market for "routine" transportation service to and from low Earth orbit and that this will be only the first step in what will be a huge opportunity for truly commercial space enterprise, inherent to the Vision for Space Exploration. I believe that the space station provides a tremendous opportunity to promote commercial space ventures that will help us meet our exploration objectives and at the same time create new jobs and new industry.

The clearly identifiable market provided by the space station is that for regular cargo delivery and return, and crew rotation, especially after we retire the shuttle in 2010 but earlier should the capability become available. We want to be able to buy these services from American industry to the fullest extent possible. We believe that when we engage the engine of competition, these services will be provided in a more cost-effective fashion than when the government has to do it. To that end, we have established a commercial crew and cargo project office and assigned to it the task of stimulating commercial enterprise in space by asking American entrepreneurs to provide innovative, cost effective commercial cargo and crew transportation services to the space station.

This fall, NASA will post a draft announcement which seeks proposals from industry for flight demonstrations to the space station of any combination of the following: external unpressurized cargo delivery and disposal; internal pressurized cargo delivery and disposal; internal pressurized cargo delivery and recovery; and crew transport.

As these capabilities are demonstrated in the years ahead, we will solicit proposals for ongoing space station transportation services from commercial providers. This announcement offers the opportunity for industry to develop capabilities that, once proven, NASA will purchase with great regularity, just as we regularly purchase launch services for our robotic spacecraft today. Once the announcement is on the street, we will receive proposals by late January with the intent to execute agreements by May of next year.

This competition will be open to emerging and established companies, with foreign content allowed, consistent with American law and policy. Proposals can include any mix of existing or new designs and hardware. NASA does not have a preferred solution. Our requirements will be couched to the maximum extent possible in terms of performance objectives, not process. Process requirements that remain will reflect matters of fundamental safety of life and property or other basic matters. It will not be government "business as usual." If those of you in industry find it to be otherwise, I expect to hear from you on the matter.

With this plan, and providing of course that we retain the support of Congress necessary to carry it out, we will put about a half-billion dollars in play over the 5 years to promote competition that is good for the private sector and good for the public interest. I'm confident that this kind of financial incentive, on different terms than are usual with NASA or indeed with any government entity, will result in the emergence of substantial commercial providers. Such successes will, in their turn, serve as a justification for even greater use of such "non-traditional" acquisition methods. As I have said in other venues, my use of the words "non-traditional" here is somewhat tongue-in-cheek because what we are talking about is completely traditional in the bulk of our economy, which is not driven by government procurement. In this larger economy, when there exist customers with specific needs and the financial resources to satisfy these needs, suppliers compete avidly to meet them. We need more of this in the space enterprise.

But as stated earlier, this is only the first step. An explicit goal of our exploration systems architecture was to provide an avenue for the creation of a substantial space economy by suitably leveraging government investment to meet its stated mission requirements. The architecture we announced in September was designed so that NASA would provide but would provide only the essential transportation elements and infrastructure to get beyond low Earth orbit. The heavy lift launchers and crew vehicles necessary to journey beyond low Earth orbit cannot, in anything like the near future, be provided by any entity other than NASA on behalf of the U.S. government. The analogy I have used elsewhere is that NASA will build the "interstate highway" that will allow us to return to the moon and to go to Mars.

We as a nation once had the systems to build this "interstate highway" leading out into the solar system. We should have retained and evolved them but we did not. So we need to rebuild them. But the "highways" themselves are not and are not supposed to be the interesting part. What is interesting are the destinations and, particularly to the point of the present discussion, the service stations, hotels and other businesses and accommodations that we will find at the "exit ramps" of our future "interstate highways" in space. It is here that a robust commercial market can develop to support our exploration goals and eventually go beyond them. I think we are at the start of something big, somewhat akin to what we saw with the personal computer 25 years ago.

To my point, NASA's exploration architecture does what it must. It fulfills the mission required by the president, according to the terms of a major speech and written policy. It does so in a fashion which some have labeled as "boring" or "lacking pizzazz," but which others have observed makes efficient use of the building blocks that we as a nation own today and in which the pieces "fit together like a fine Swiss watch." I believe these seemingly divergent views are merely two sides are the same coin, reflecting the fact that the plan delivers what it must without including what it need not. Nothing else is acceptable in these fiscally challenging times.

But the building blocks of our architecture could easily be used to accomplish much more, with the right leverage from commercial providers. To see how this is so, observe first that our "1.5 launch solution" separates the smaller crew launch from that of the heavy, high-value cargo, both on shuttle-derived launch vehicle variants. While this approach allows us to meet lunar return mission requirements with U.S. government systems—no external entities are in the critical path for mission accomplishment—it does not exclude such entities and indeed provides several "hooks" and "scars" by which their services can be used to facilitate or enhance the mission.

By the time we are ready to return to the moon, the space station will have been completed and will be in receipt of routine commercial resupply and crew rotation service for, we hope, several years. So, if the plan for stimulating the development of space station commercial crew rotation capability is successful, it becomes possible to envision the crew launch phase of the lunar mission being carried out on commercial systems. This would be a service we could purchase commercially, leaving the very heavy lift requirements to the government system, for which it is less likely that there will be other commercial applications during this period.

Whether or not this occurs, other options are also possible. Astute observers will note that the shuttle-derived heavy lift vehicle (SDHLV) that we have proposed is not, as a rocket, being optimally utilized for its lunar mission. This is because some of the fuel in the so-called "Earth departure stage" is used to lift the lunar payload into Earth orbit, but additional fuel must yet be retained for the translunar ignition burn of over 3 kilometers per second. From a purely architectural point of view, the SDHLV is an expensive vehicle most aptly utilized for lifting only expensive cargo such as the man-rated systems it carries. But in our architecture, some of its lift capacity must be utilized to carry fuel into low Earth orbit. This is unsatisfying, because when on the ground, fuel is about the cheapest material employed in any aspect of the space business. Its value in orbit (at least several thousand dollars per pound) is almost completely a function of its location rather than intrinsic to its nature. In contrast, the value of, say, the Lunar Surface Access Module (LSAM) brought up on the heavy-lifter will be well over $100 thousand per pound, most of which represents its intrinsic cost. The additional value it acquires when transported to its new position in low Earth orbit remains a small part of the total value.

Logically, then, we should seek to use the SDHLV only for the highest-value cargo, and specifically we should desire to place fuel in orbit by the cheapest means possible in whatever manner this can be accomplished, whether of high reliability or not. However, in deciding to embark on a lunar mission, we cannot afford the consequential damage of not having fuel available when needed. Recognizing that fact, our mission architecture hauls its own Earth-departure fuel up from the ground for each trip. But if there were a fuel depot available on orbit, one capable of being replenished at any time, the Earth departure stage could after refueling carry significantly more payload to the moon, maximizing the utility of the inherently expensive SDHLV for carrying high-value cargo.

But NASA's architecture does not feature a fuel depot. Even if it could be afforded within the budget constraints (which we will likely face) and it cannot; it is philosophically the wrong thing for the government to be doing. It is not "necessary;" it is not on the critical path of things we "must do" to return astronauts to the moon. It is a highly valuable enhancement, but the mission is not hostage to its availability. It is exactly the type of enterprise which should be left to industry and to the marketplace.

So let us look forward 10 or more years to a time when we are closer to resuming human exploration of the moon. The value of such a commercially operated fuel depot in low Earth orbit at that time is easy to estimate. Such a depot would support at least two planned missions to the moon each year. The architecture which we have advanced places about 150 metric tons in low Earth orbit, 25 metric tons on the Crew Launch Vehicle and 125 metric tons on the heavy-lifter. Of the total, about half will be propellant in the form of liquid oxygen and hydrogen, required for the translunar injection to the moon. If the Earth departure stage could be refueled on-orbit, the crew and all high-value hardware could be launched using a single SDHLV; and all of this could be sent to the moon.

There are several ways in which the value of this extra capability might be calculated; but at a conservatively low government price of $10,000 per a kilogram for payload in low Earth orbit, 250 megatons of fuel for two missions per year is worth $2.5 billion at government rates. If a commercial provider can supply fuel at a lower cost, both the government and the contractor will benefit. This is a non-trivial market, and it will only grow as we continue to fly. The value of fuel for a single Mars mission may be several billion dollars by itself. Once industry becomes fully convinced that the United States in company with its international partners is headed out into the solar system for good, I believe that the economics of such a business will attract multiple competitors, to the benefit of both stockholders and taxpayers.

Best of all, such an approach enables us to leverage the value of the government system without putting commercial fuel deliveries in the critical path. If the depot is there and is full, we can use it. But with the architecture we have advanced, we can conduct missions to the moon without it. The government does not need to have oversight, or even insight, into the quality and reliability of the fuel delivery service. If fuel is not delivered, the loss belongs to the operator, not to the government. If fuel is delivered and maintained in storage, the contractors are paid, whether or not the government flies its intended missions. If long-term delivery contracts are negotiated, and the provider learns to effect deliveries more efficiently, the gain is his not the government's. Since fuel is completely fungible, it can be left to the provider to determine the optimum origin, size and method of a delivering it. And finally, though I would rather not do it, it is even possible that we could develop such a market in stages, with the first fuel tank provided by the government and then turned over to a commercial provider to store and maintain fuel for future missions, and to expand the tank farm as warranted by the market.

To maintain and operate the fuel depot, periodic human support may be needed. Living space in Earth orbit may be required; if so, this presents yet another commercial opportunity for people like Bob Bigelow, who is already working on developing space habitats. So the logistics needs of the fuel depot may provide more of the same opportunities that we will pioneer with the space station.

Fuel and other consumables will not always be most needed where they are stored. Will orbital transfer and delivery services develop, with reusable "space tugs" ferrying goods from centralized stockpiles to other locations?

The fuel depot operator will need power for refrigeration and other support systems. This might well be left to specialty suppliers who know nothing of the storage and maintenance of cryogenic tank farms, but who know a lot about how to generate and store power. Could these be standard power modules, developed and delivered for a fee to locations specified by the user?

In the course of conducting many fuel replenishment missions and associated operations, commercial launch and orbital systems of known and presumably high reliability will be developed and evolved. Government mission planners will be able to take advantage of these systems, which will become "known quantities" by virtue of their track record rather than through the at best mixed blessings of government development oversight.

There will also be a private sector role in supporting a variety of lunar surface systems and infrastructure, including lunar habitats, power and science facilities, surface rovers, logistics and resupply, communications, navigation and *in situ* resource utilization equipment. There may or may not be gold on the moon—I'm not sure we care—but we may well witness a 21st century gold rush of sorts when entrepreneurs learn to roast oxygen from the lunar soil, saving a major portion of the cost of bringing fuel to the lunar surface. Will a time come when it is more economical to ship liquid oxygen from the lunar surface to low Earth orbit then to bring it up from Earth?

This will all start to become "really real" in 10 years or so. As I see it, these are exactly the kinds of enterprises to which government is poorly suited, but which in the hands of the right entrepreneur can earn that person a cover on *Fortune* magazine. But it will take enlightened government management to bring it about, management as much in the form of "what not to do" as "to do." In the coming years and decades, NASA must focus on its core government role as a provider of infrastructure broadly applicable to the common good and too expensive for any single business entity to develop. NASA must remain on the frontier and must conscientiously architect its plans to favor the inclusion of entrepreneurs through arms-length transactions wherever possible, restricting the use of classic "prime contracts" to situations where they are the right tool and not the default tool.

With the beginning of space station operations 5 years ago, we are now at a point children born at the beginning of the 21st century will live their lives knowing that there will always be people living and working in space. And the number of people who will be engaged in such activity will grow by leaps and bounds if we in government are faithful in executing our role in helping the private sector to step up to these new opportunities. I hope there are many entrepreneurs in this audience who have the vision to help us help them pioneer the commercial space frontier. You, and all those engaged in the quest that we are undertaking, have my sincere thanks and appreciation.

American Competitiveness: NASA's Role & Everyone's Responsibility

Michael D. Griffin
Administrator
National Aeronautics and Space Administration
Lecture for The January Series at Calvin College
January 17, 2008

I'd like to start by recalling a congressional hearing with the late Physicist Robert Wilson, co-discoverer of the 3 degree kelvin microwave background radiation that is the remnant of the 14 billion year old Big Bang. When asked before a committee about what value a new particle accelerator would have in promoting the national security of our country, he responded: "Nothing at all. It only has to do with the respect with which we regard one another, the dignity of men, our love of culture. ... It has to do with are we good painters, good sculptors, great poets? I mean all the things we really venerate in our country and are patriotic about. ... It has nothing to do directly with defending our country except to make it worth defending."

Similarly, NASA's scientific activities in climate change research, monitoring our ever-changing sun and studying the physics of solar flares and their effects on our Earth, our missions to the other planets, moons, asteroids and comets of our solar system as well as our astronomy and astrophysics missions, like the Hubble Space Telescope, make our country worth defending. Further, I hope that the space shuttle, International Space Station and our next missions to the moon (this time to stay) are something of which we are all proud. These are the things that make our nation worth defending.

I recently read an essay written a few years ago by Michael Crichton, the author of many popular science fiction books, including *Jurassic Park* and *The Andromeda Strain*. In that article, Crichton highlighted the work of a privately funded foundation called Space Camp, an intensive program for kids and adults to teach the physics and engineering of spaceflight. Last year, after 25 years of operation, Space Camp graduated its 500,000th camper. In his essay, Crichton tells the story of a 10 year-old boy who was interviewed on TV after graduating from Space Camp. "Asked about the future, he spoke of colonies on the moon, and trips to Mars. The reporter said, 'How are you going to get Congress to pay for it?'" To which the young boy replied, "Maybe your Congress won't, but mine will." With your help, with American ingenuity and support, we are slowly turning this young man's dreams into reality.

At a fundamental level, NASA is in the inspiration business. We're about making our country worth defending and I am extremely lucky and proud to be a part of this great enterprise.

It invigorates me to visit a college campus and meet the next generation of physicists and engineers, to hear about the latest research they are conducting and to meet the young people who will go on to build our nation's new spacecraft and launch vehicles and discover new things about our Earth, solar system and universe; or build our nation's next generation air traffic control system; or design advanced aircraft to make air travel safer, cheaper, faster and more environmentally friendly. I just met with some of the future professional engineers and scientists of Calvin College this morning and, as always, I really enjoyed the Q&A.

But the questions make me realize, increasingly, that I am two generations removed from the life and world of undergraduate education. And, sometimes, I am told that young people today are just not interested in NASA, in the space program, and that my generation cannot possibly understand the college students of today. After all, I grew up in the very different world of the 1950s and 60s. Today, we have satellite television with hundreds of channels and 24-hour news coverage, inexpensive jet travel, personal computers, cell phones and instant messaging, et cetera; so how could I possibly understand this new generation? Now, I will readily admit to being clueless about a lot of popular culture, but despite that, I think the best answer I can give is, "You're right. My generation didn't have all those things when I was young. We invented them."

Now, some of you in this auditorium are of my generation, which grew up during the Apollo era of the 1960s, NASA's apotheosis. We watched science fiction movies and television shows that made us believe that we—all of us and not simply a few astronauts—could become space travelers. Arthur C. Clarke's and Stanley Kubrik's masterpieces of science fiction *2001: A Space Odyssey* projected onto the screen of our collective human consciousness a future for us where, by now, hundreds of people would be living and working in space stations orbiting Earth and towns would exist on the moon. We would be journeying to other planets in our solar system, just as our European forebears came to America looking for new beginnings. This vision of our future proved illusory for our generation for two fundamental reasons: the limitations of our economic resources and of our technology. Neil Armstrong's "giant leap for mankind" was not a journey that could be sustained without a more concerted investment of time, resources and energy than the nation was willing to provide after July 20, 1969.

But rather than looking back wistfully on past greatness, I would rather learn from such history to understand our present and predict our future in space exploration. NASA celebrates it 50th birthday this year, but that does not mean we are due for a midlife crisis; it means that we have reached a milestone to recognize, celebrate and then blow out the birthday candles with the wish that we be refreshed and renewed in our approach to the problems we face today and are likely to face in the future.

We have been exploring space now for 50 years; but it has only been 50 years. By way of comparison, human beings have been conducting transoceanic voyages for 1,000 years or so. So, in only the first 50 years of spaceflight, it is actually quite remarkable to realize that NASA's robotic spacecraft have ventured to almost all planets in the solar system; four have actually left the solar system;

and 12 men have walked on the moon. We are in the midst of constructing the space station, which will be larger in wingspan than a football field and weigh about what the first Mars ship will weigh. Its development is the largest task ever performed by the civilian agencies of the United States or our international partners; only military coalitions have undertaken larger efforts.

Yet despite the achievements of our nation's first 50 years in space, the history books 1,000 years from now will note that the United States of America was not the first country to explore space. Those books will name a nation that no longer exists—the Union of Soviet Socialist Republics. Those books will show that the Soviet Union launched the first man-made object into space, *Sputnik*, in October 1957; and that they launched the first astronaut, Yuri Gagarin, in April 1961. I was a young boy, 8 years old at the time of *Sputnik,* growing up around an Army base in Aberdeen, Maryland, and I can still remember vividly the fear and embarrassment our nation felt at that time. It was on the front page of every newspaper, in the largest possible type-font. The idea that the United States could be beaten to space by any other nation, not to mention by our supposedly-backward declared adversary, was for almost everyone a galvanizing event. Nikita Khrushchev's November 1956 admonition—"We will bury you"—reverberated in our collective consciousness. *Sputnik* shifted the arena of international technical competition to the new frontier of space and it mattered greatly.

One of the national leaders who recognized the importance of *Sputnik* was a young congressman from Grand Rapids by the name of Gerald Ford who in 1958 volunteered to become a member of the House Select Committee on Astronautics and Space Exploration. This committee has in the course of 50 years evolved into the House Science and Technology Committee. More importantly, this congressional committee and Congressman Ford in particular, was important in the drafting of the original Space Act legislation which founded NASA, bringing together laboratories and field centers from various other branches of the federal government, including the Army, Navy and the civilian National Advisory Committee on Aeronautics.

When Gerald Ford became president 16 years later, he saluted the landings of the twin Viking robotic explorers on Mars, saying on the occasion of the first landing, "Our achievements in space represent not only the height of technological skill, they also reflect the best in our country—our character, the capacity for creativity and sacrifice and a willingness to reach into the unknown." In the summer of 1975, President Ford also spoke via telephone through NASA ground antennas to American astronauts Tom Stafford and Deke Slayton and Soviet cosmonaut Valeriy Kubasov onboard the Apollo-*Soyuz* spacecraft 140 miles overhead. In the span of only a few years, America went from being behind in the space race to putting 12 men on the surface of the moon. We also went from a competition to the beginning of a partnership with the Soviet Union; and our collaboration continues to this day. Partnership with other spacefaring nations has become a vital element of the United States "soft power" appeal. And over half of all NASA science missions, with over 50 spacecraft operating in space today, involve some form of international collaboration.

Today, 200 miles overhead on the space station, NASA astronauts Peggy Whitson and Dan Tani are living and working in space with Russian cosmonaut Yuri Malechenko. With the space station, NASA and our 15 international partners have maintained a permanent human foothold in space since October 2000—over 7 years, and we are still learning the hard lessons of how to live and work in space 24/7/365. We are in the midst of space station assembly with the space shuttle between now and the fall of 2010 and hope to launch the European *Columbus* module in 2 weeks with Space Shuttle *Atlantis*, commanded by Navy Commander Steve Frick. *Atlantis* will also deliver German astronaut Hans Schlegel as part of the assembly team, and leave French astronaut Leopold Eyharts on the space station, replacing U.S. astronaut Dan Tani.

We are using the station as a laboratory test bed for technologies, techniques and lessons that will enable future colonies on the moon and trips to Mars; and we are also developing materials and conducting research, which will benefit us here on Earth. For example, Peggy and Dan recently activated a Microgravity Science Glovebox experiment called InSpace. The purpose of this investigation is to obtain fundamental data of the complex properties of a class of smart materials termed magnetorheological (MR) fluids. Magnetorheological fluids are suspensions of small (micron-sized) superparamagnetic particles in a nonmagnetic medium. These controllable fluids can quickly transition into a nearly solid-like state when exposed to a magnetic field and return to their original liquid state when the magnetic field is removed. The relative stiffness can be controlled by controlling the strength of the magnetic field. Thus, due to the rapid-response interface that they provide between mechanical components and electronic controls, MR fluids can be used to improve or develop new brake systems, seat suspensions, robotics, clutches, airplane landing gear and vibration damping systems.

Last year, a convention of the American Medical Association (AMA) endorsed NASA's efforts in human spaceflight, in going to the moon, Mars and beyond because the technologies and techniques we have developed for doctors will "undoubtedly yield both projected and unanticipated biomedical breakthroughs." The AMA resolution listed several NASA contributions to their work, including LASIK surgery, laser angioplasty, dialysis machines improvements and digital cochlear implants.

One of the success stories from NASA's work to develop such countermeasures is against painful kidney stones. In microgravity, the human body compensates for the reduced stress on the skeleton by releasing calcium from our bones, making astronauts more prone to developing kidney stones. In order to prevent the formation of such stones, astronauts have been taking potassium citrate and NASA is conducting experiments with a new generation of pharmaceuticals with companies like Amgen to test other ways to prevent or reduce osteoporosis-like bone loss as well as deteriorating muscles.

Last September, Elias Zerhouni of the National Institutes of Health (NIH) and I signed a Memorandum of Understanding to conduct even more joint medical research onboard the space station. On the next shuttle flight, STS-122, NASA astronauts will test a drug called midodrine with the help of NIH researchers to hopefully reduce dizziness caused by a drop in blood pressure after our astronauts first return back to Earth from the zero-g environment of space.

Again, our goal is to develop and test new capabilities onboard the space station that cannot be tested anywhere on Earth; and that will not only enable future spaceflight missions to the moon, Mars and beyond but also benefit life here on Earth.

NASA simply cannot carry out this ambitious goal of exploring the solar system alone. We will need international collaborators, commercial companies, venture capitalists and other agencies of the United States government. It will take American know-how and can-do attitude. It will literally take "the best of the best of the best" to turn this goal into a reality. In my usual clueless fashion, I had failed to notice—until receiving a question from a member of the media—that Peggy Whitson is the first woman to command the space station. Peggy has a Ph.D. in biochemistry, studying at Iowa Wesleyan University College and Rice University in Houston, Texas. She is a veteran astronaut, who previously lived and worked for 6 months onboard the station in 2002 as the science officer. And yes, NASA's naming convention here with "science officer" pays homage to Star Trek's Mr. Spock. However, pointy ears are not required for this job.

Peggy is literally one of "the best of the best of the best," because less than 1 percent of those who even apply to become astronauts are selected. Over the years, NASA has received approximately 41,000 applications from prospective astronauts, while only 321 individuals have been selected. NASA is in the process of taking applications and screening for the next class of astronaut candidates even now, and we plan to announce this selection early next year.

Times have changed from the NASA of the 1950s and 1960s; and they should. The stereotypical buzz-cut test pilot or white male engineer like me is no longer representative of our agency. NASA depends upon the ideas in our people's heads for our success, not upon the package containing them. And while I do indeed care about the egalitarianism of society, I am also being pragmatic. For America to continue to be pre-eminent in the world economy, to be the world's leader in innovation, science and technology and to be a leader on the frontier of space exploration and aeronautics research, NASA will need the best ideas, hard work and dedication from all those who would like to be involved with this most exciting enterprise of our time.

To explore space, we will need people, energy and resources, so let me address the facts and some common misconceptions about how much the American taxpayer provides for NASA's budget. America's annual investment in NASA is less than one penny out of every federal dollar spent. Let me repeat: If you looked into your wallet or purse and pulled out a dollar bill and a penny, the entire federal budget represents that dollar while NASA's budget is less than that penny. To be more exact, NASA's current budget is six tenths of 1 percent of every federal dollar spent. This is somewhere in the realm of what engineers like me call rounding error. However, when polled, the average American believes NASA's budget to be much higher than it actually is, 24 percent of the federal budget, comparable to that of the Pentagon. In fact, NASA's budget this year is $17.3 billion, the Pentagon's operating budget (not including supplemental appropriations for our operations in Iraq and Afghanistan) is $459 billion and the overall federal budget is over $2.5 trillion.

From this small investment in NASA over many years, new engineering and scientific capabilities built originally for our nation's space program are now pervasive in our lives, critical to a range of activities that create and provide value. Since the 1960s, NASA pioneered research in high bandwidth satellite communications which helped lead to the development of high-definition satellite television with 24-hour news, entertainment and sports anywhere in the world.

Forty years ago, engineers like me used three pieces of wood and a piece of plastic—the slide rule—to make calculations. Thirty years ago, 1,000 transistors could fit on a silicon chip; today, it's 1.7 billion. The cost of such chips has dropped by a factor of 100,000. Few people know that the development of the first microprocessors was born of a competition between Fairchild and Intel in the 1960s to build components small enough to fit in NASA's Apollo spacecraft.

We built weather and climate change monitoring sensors and satellites that, along with the fundamental research and applications from this data, improve our daily lives. Working with the Air Force and Navy, NASA improved precision timing techniques with atomic clocks that enabled the development of the Global Positioning System (GPS), which created a consumer market of over $20 billion in sales this year. In every GPS satellite, there is a small correction to its atomic clock to compensate for the effects of special and general relativity discovered by Albert Einstein.

In partnership with the Federal Aviation Administration (FAA), NASA is developing the concepts, algorithms and technologies to increase the airspace capacity in the United States in a safe, equitable and efficient manner. A key question here is how to best address where, when, how and the extent to which machine-level automation of air traffic control functions can be safely and effectively applied throughout U.S. airspace. NASA is also not limiting its research simply to the airspace; we are also looking at ways to improve the efficiency in the use of airport gates, taxiways and runways while balancing the requirements of safety and environmental concerns. Researchers from across the United States have used NASA's aerodynamics laboratories, wind tunnels and know-how to help develop every single jet fighter aircraft used by the Air Force and Navy and to test new, commercial jet engines and lightweight composite structures.

Again, my generation didn't have these things when I was young. We invented them. Sometimes our contribution is not to create new technologies but to integrate various existing capabilities in innovative ways. Last fall, NASA used its air and space capabilities to aid Californians during the terrible wildfires that ravaged Southern California. Our Earth-observing satellites helped monitor the spread of those terrible fires. We also sent an unmanned aerial vehicle (UAV) equipped with unique infrared (IR) sensors to fly over the fires. The Ikhana UAV, which is operated through a cooperative effort between the Ames and Dryden Research Centers in California, peered through heavy smoke and darkness, found hot spots and flames and transmitted the sensor information to a computer server at Ames where it was combined with Google Earth maps and then transmitted to operations centers to provide firefighters a much better understanding of the situation, aiding disaster managers in allocating firefighting resources. The quick turnaround made a difference too. Information gathered from piloted airplanes currently must wait for the aircraft to land before it can be transmitted, while the Ikhana UAV sent the data to fire incident commanders only minutes after acquisition. Eventually and in concert with other agencies, we at NASA hope to have an entire network of sensors that will provide information about natural disasters at every scale, from the ground up to space, aiding responders and hopefully saving lives.

In another example, NASA is helping the poor countries of Central America with SERVIR (Spanish for "to serve"), a high-tech satellite visualization system that monitors weather and climate, helps to track and combat wildfires, improves land use for city planning and agricultural practices and helps local officials respond faster to natural disasters. Meteorologists and disaster response experts in Central America use SERVIR to see where rain will fall, where flooding will occur, the location of forest fires, hurricanes, tornadoes and pretty much anything nature can dish out. Most recently, NASA research brought together radar imagery and other satellite data to help the Dominican Republic's government respond to extensive flooding in the wake of Tropical Storm Noel. The SERVIR project along with other acts of kindness and charity by the embedded NASA team has been such a success that one of our researchers, Dan Irwin, actually found himself being nominated to be the mayor for the small town of San Andres, Guatemala. Dan respectfully declined, but he was touched by the vote of confidence. NASA is now working with the State Department, the National Oceanic and Atmospheric Administration (NOAA) and other agencies to help provide capabilities like SERVIR to other regions of the world, like Africa.

Again, NASA is bringing space capabilities to bear to improve people's lives and even to save lives. But it will take far more than NASA funding to open up the new, exciting opportunities we hope to continue finding when we explore and exploit the vantage point of space.

NASA has formed a strategic partnership with the founders of Google to carry out various scientific endeavors, like the Google Moon mapping software, the use of their Gulfstream V to carry out scientific missions such as the campaign to monitor the Quadrantid meteor shower earlier this month, and supporting Google's offer of a prize purse of up to $30 million for the first privately funded and developed lander/rover to touch down successfully on the moon and carry out various experiments. I also hope to open up the space station as a national laboratory to commercial ventures and create relationships similar to our work with Amgen and other pharmaceutical companies.

My hope is that more people will be able to experience and benefit from space exploration and scientific discovery and even make a profit from it. That is the American way. Likewise, it is also my hope that NASA will be able to spur on and leverage the capabilities that the commercial sector builds and be able to harness the improved intellectual capabilities coming from our nation's universities and high school students. This is important. It matters greatly to our nation's future.

I am gravely concerned when I read statistics about how, on average, U.S. students are lagging behind their counterparts in other countries in their knowledge of math and science. According to a recent report which measures the skills of 15-year-olds in math and science across 30 industrialized nations, American students are trailing many potential competitors and sometimes trailing badly. On average, U.S. students placed below standards in science, well behind Japan and Korea, but also trailed Ireland and Iceland. American 15-year olds did even worse in math, trailing many nations in Asia and Europe.

These troubling trends were best explored by the recent report *Rising Above the Gathering Storm* by the National Academy of Engineering. One of the first paragraphs in the report captured the situation well, so I will quote it at length: "Having reviewed the trends in the United States and abroad, the committee is deeply concerned that the scientific and technical building blocks of our economic leadership are eroding at a time when many other nations are gathering strength. We strongly believe that a worldwide strengthening will benefit the world's economy—particularly in the creation of jobs in countries that are far less well off than the United States—but we are worried about the future prosperity of the United States. Although many people assume that the United States will always be a world leader in science and technology, this may not continue to be the case inasmuch as great minds and great ideas exist throughout the world. We fear the abruptness with which such a lead in science and technology can be lost and the difficulty of recovering a lead once lost—if indeed it can be regained at all."

This is a sobering assessment. This report also cites some alarming statistics. Fifty years ago, almost twice as many bachelor's degrees in physics were awarded in the United States than in 2004. Last year, the United States produced more undergraduates in sports exercise than in electrical engineering. About a third of U.S. students who plan to study engineering when they entered college switch majors before graduating; they probably are not switching to mathematics or theoretical physics. Today, there are more software engineers in Bangalore, India than in Silicon Valley. In 2000, 38 percent of all U.S. science and technology Ph.D.s were conferred upon foreign-born graduate students, most of whom return to their home countries.

I hope you agree with me that America's economic growth is driven by technological innovation and that societies which foster such innovation become leaders in the world. So, as NASA begins its next 50 years, I am deeply concerned about our nation's "bench strength" in carrying out our mission of space exploration, as well as other technical endeavors. We still need "the best of the best of the best" in more than just the astronaut corps. This is rocket science. The alarming statistics I have quoted have broad implications for the United States' ability to maintain economic and technological leadership in today's world.

Specific to the realm of spaceflight, I am concerned that America's real and perceived leadership in the standing of the world's spacefaring nations is slipping away. As Admiral Hal Gehman noted in his report of the Space Shuttle *Columbia* Accident Investigation Board a few years ago, "previous attempts to develop a replacement vehicle for the aging shuttle represent a failure of national leadership."

That is also a sobering assessment. We have only recently begun developing the new Orion Crew Exploration Vehicle and Ares rockets, which will ferry astronauts to and from the space station and, more importantly, allow us once again to go beyond low Earth orbit to the moon. We plan to retire the space shuttle in 2010, after nearly 30 years of experimental flights. However, with current budget projections, NASA's new human spaceflight systems will not come online until 2015. With an operational stand-down like this, I am gravely concerned that even more highly-skilled engineers will simply exit the field altogether, as happened at the end of the Apollo program. Worse, between now and then, NASA will pay over $700 million (and possibly a good deal more) to the Russian Space Agency to support the space station with their *Soyuz* and *Progress* crew and cargo vehicles. Other countries, like Malaysia and South Korea, and certain wealthy individuals are already paying the Russians for trips to the space station. So, 50 years after *Sputnik* and 35 years after the last American footprint on the moon, I must ask the question: who is currently the recognized leader in spaceflight?

China has also emerged as a major spacefaring nation. China demonstrated an antisatellite weapon against one of its own aging weather satellites a year ago and launched its first satellite mission to the moon last October. In 2008, the Chinese plan to launch 17 satellites and to conduct their first spacewalk following the Beijing Olympics this fall. China is investing heavily in building its space capabilities because it understands the value of these activities as a driver for innovation and a source of national pride in being a member of the world's most exclusive club. China today not only flies its own taikonauts but also has plans to launch about 100 satellites over the next 5 to 8 years. It should be no surprise, especially to those who have read Tom Friedman's book *The World is Flat* or John Kao's *Innovation Nation*, that this environment in China is breeding thousands of high-tech startups.

The Chinese have adapted the design of the Russian *Soyuz* to create their *Shenzhou* spacecraft. However, the similarity between the two ends at the outer mould line; the *Shenzhou* spacecraft is both more spacious and more capable. They plan to conduct their first spacewalks and orbital rendezvous operations and to build their own space station (admittedly simpler than ours) in the coming years. While they have not stated an intention to do so, the Chinese could send a mission around the moon with the *Shenzhou* spacecraft, as the United States did with the inspiring Apollo 8 mission back in 1968. China could easily execute such a mission with its planned Long March V rocket currently under development and reportedly rivaling the capabilities of any expendable rocket in the world today. I have no doubt that they will have it in use, as they plan, by around 2012.

I am pointing out such things, matters of engineering capability, because I believe that it is important to understand our strategic competitors as well as those with whom we wish to collaborate. We must also understand ourselves and the framework of our real and perceived leadership in the world in a broader context than simply NASA's six tenths of 1 percent of the federal budget. As John Kao couches the issue, we are currently facing a "Silent *Sputnik*" where "many countries are racing for a new innovation high ground while our own advantages are showing signs of serious wear."

If you agree with me that our nation is indeed facing a "Silent *Sputnik*" moment, then this situation begs the question: why does it take a crisis to get our nation's attention? I am concerned that America's potential as a great nation is withering away due to benign neglect, apathy, complacency and a lack of leadership. That is, we are ignoring the crisis because there is not a galvanizing moment like the launch of *Sputnik*.

Now I fully appreciate there are many distractions in our modern life today, possibly due to the 24-hour satellite news capabilities that NASA itself helped to create. Last summer, just prior to a space shuttle launch, I sat down for an interview with CNN just as one of their producers informed me that they had to cut away from their coverage of the shuttle launch. There was breaking news of vital national interest from Los Angeles: Paris Hilton was going to jail. And NASA could not compete for the American people's attention against Paris Hilton. That was the moment when I realized how tough the NASA administrator's job really is.

While I make light of this, there is a not-so-subtle lesson here, that our media and nation are not focusing enough on what matters most. Thus, I believe it is necessary for us—all of us—to discuss openly the founding principles that led us as a nation to embrace space exploration five decades ago.

A former chairman of the House Science Committee, Congressman Bob Walker from Pennsylvania, framed the issue very well in a speech soon after the Space Shuttle *Columbia* tragedy 5 years ago: "For every generation, choices are made that lead to greatness or mediocrity." And I would ask that all of us, each and every one of us here today, consider our choices and decisions we make in how we spend our time, resources and energy.

In this thought-provoking speech, Congressman Walker quoted from the great British statesman, Benjamin Disraeli, who once opined that "nations go from bondage to faith, from faith to courage, from courage to freedom, from freedom to abundance, from abundance to complacency, from complacency to dependency and from dependency back to bondage." It's all a matter of what each generation, in its time here on Earth, chooses to do.

History books hundreds of years from now will note President John F. Kennedy's choice for America in 1962. "We choose to go to the moon in this decade and do the other things not because they are easy but because they are hard; because that goal will serve to organize and measure the best of our energies and skills; because that challenge is one that we are willing to accept, one we are unwilling to postpone and one which we intend to win, and the others, too. It is for these reasons that I regard the decision last year to shift our efforts in space from low to high gear as among the most important decisions that will be made during my incumbency in the office of the presidency."

When President Kennedy spoke those bold words and challenged our nation, NASA then had less than 11 hours of experience in human spaceflight under its belt in the Mercury program; but we had "The Right Stuff." We did not yet have the Apollo capsules or powerful Saturn V rockets or lunar landers;

we did not even have computers as advanced as the Blackberry I have on me today, let alone the power of the Internet. We invented them.

"For every generation, choices are made that lead to greatness or mediocrity." Thank you for choosing to spend this afternoon listening to me.

Space Exploration: A Frontier for American Collaboration

Michael D. Griffin
Administrator
National Aeronautics and Space Administration
Loewy Lecture
Georgetown University
November 16, 2007

A few weeks ago, I was asked to speak about the role of space exploration in spurring innovation and American competitiveness in the world. Today, I would like to address the opposite question: how can space exploration spur greater collaboration between our nation and others?

It is necessary to be successful both in competition and in collaboration if we are to survive and prosper, whether as individuals or as a society. We cannot thrive if our presence offers nothing to others that they cannot do more easily themselves. And we cannot thrive if every other hand is turned against us. So, I believe that it is important to strike a thoughtful balance between competition and collaboration. In the most fundamental sense of these words, it is crucial to our national security to do so.

"National security" is an elusive concept and its fulfillment imposes different requirements upon a great nation than upon a small one. Most obviously, it consists of having the wherewithal to act, militarily, in support of our nation's perceived interests. At a higher level, it consists in part of a measure of deterrence against potential adversaries. In George Washington's famous words, "if you would have peace, prepare for war." But I would submit to you that the highest possible form of national security, well above having better guns and bombs, is that which comes from being a nation which seeks to carry out the great deeds that cause other countries to want to join with us in pursuing those objectives. In this sense, it is of enormous value to our nation to collaborate with others in the most technically challenging endeavor of our time—space exploration.

As the present administrator of NASA, I am fortunate to bear witness to an enormous effort carried out daily on the frontiers of both technology and international cooperation. With 16 participating nations, the International Space Station under construction today is a testament to the perseverance of the United States, Russia, the countries of the European Space Agency, Japan and Canada working together on the largest task ever performed by the civilian agencies of the United States or our international partners. On November 2, we celebrated 7 years of permanent human presence in space onboard the space station. The partnership that brought it about has endured tremendous hardships, most especially the loss of Space Shuttle *Columbia*, and stands by itself as a monumental international accomplishment. The space station will indeed pay dividends as an engineering and research laboratory as we push outward in Constellation, the successor to Apollo, back to the moon and then on to Mars and other destinations in our solar system over the course of the next decades. But eventually, the space station hardware will fail or the questions we can pose with it will have been asked and answered. Eventually, and so that it does not become a danger, it will be reentered into the Pacific.

Thus, one day the space station will be no more. But I believe that one day we will conclude that most important legacy of the space station endeavor was the partnership itself. Together, we are learning the hard but essential lessons concerning how we can carry out the largest and most complex endeavors human beings have yet undertaken.

I do not say this lightly. The station rivals the Apollo program in cost, and in my opinion easily surpasses it in complexity. When completed, it will be longer than a football field, four times larger than the Russian *Mir* space station and five times larger than the 1970s *Skylab*. It is truly one of the great engineering wonders of the world akin to such feats as the Great Wall of China, the pyramids of Egypt, the Panama and Suez canals or the sea walls of Venice.

Last month, Space Shuttle *Discovery* delivered the Italian-built *Harmony* Module to the space station along with Italian astronaut Paolo Nespoli as part of the assembly team. It was the most challenging space station mission undertaken thus far and it was completed brilliantly, including for good measure a contingency spacewalk to effect repairs to a torn solar array. Next month, Space Shuttle *Atlantis* will launch the European *Columbus* Laboratory Module, assembled in Bremen, Germany. Space Shuttle *Atlantis* will also deliver German astronaut Hans Schlegel as part of the assembly team and leave French astronaut Leopold Eyharts on the space station, replacing U.S. astronaut Dan Tani.

Independent human spaceflight has been accomplished only by the United States, Russia and China. India has announced its intention to develop such capabilities, joining this most exclusive club of spacefaring nations. Having visited several space facilities in China and India last year, and having met their aerospace engineers, I must say that I am very impressed by the methodical, disciplined approach that these nations have taken in developing their space industrial base and capabilities. The national economies of both countries exceed in scale the economy of the United States as it existed in the early 1960s when America set out to undertake the Apollo program in accordance with President John F. Kennedy's vision for our nation's future on the "New Frontier." So if they wish to send their own astronauts into space, it is simply a matter of national will on their part, of choosing to do so.

But rather than fostering a new rivalry in space I hope that China and other countries will join their own programs to the United States' effort in Constellation, returning together to the moon and exploring space to our mutual benefit. In this regard, China's anti-satellite weapon demonstration last January was a step backward. We can all hope that it will be the only such step.

Last September, Japan launched the *Selene* mission to Earth's moon and NASA has an agreement with the Japanese Space Agency to share the data collected from that mission. China also launched its first lunar mission, *Chang'e*, last month. I want to applaud China's recent announcement that it would provide the data collected from this mission to researchers around the world in accordance with common international practice. We recently established a new Lunar Science Institute at the Ames Research Center in the heart of Silicon Valley, California. Our goals with this institute are to use state-of-the-art information technologies, like Google's recent partnerships with NASA, to create new virtual and international collaborations for lunar research and to spark the growth of a lunar science community.

We will use the data collected from these spacecraft, from India's *Chandrayaan* as well as NASA's Lunar Reconnaissance Orbiter and Lunar Crater Observation and Sensing Satellite missions, all planned for launch next year, to produce a detailed map of the lunar surface and its resources as well as to better understand its gravity field, to search for evidence of polar volatiles and to define radiation hazards so we can mitigate them for human missions beginning in the next decade. We will need such data to carry out our nation's plans to build our first outpost on the new frontier of the moon.

We are actively seeking out other countries in this journey to explore the undiscovered country of our moon and other worlds. Today, over half of our 50-plus operating robotic science missions incorporate some form of international collaboration. These include a wide range of missions to other planets and moons in our solar system as well as comets and asteroids. They include Earth science missions enabling the study of climate change by a community of international researchers for which NASA is, by far, the greatest contributor. And they include heliophysics missions like Ulysses and the Solar and Heliospheric Observatory to help us to understand our own sun and (of course) great astrophysical observatories like the Hubble Space Telescope.

Space exploration, whether human or robotic, is the grandest and most technically challenging expression of human imagination of which I can conceive. Throughout my professional career, I've wanted nothing more than to be a part of it. And I think it is in our nation's best interests to work together in this unique human endeavor, to learn from each other, as different countries and cultures, how we go about solving the unique problems presented by the exploration of space. My training in physics tells me that the problems and constraints are the same for all; the rocket equation does not change when expressed in another language. But my training and experience as an engineer has taught me that the vagaries of human ingenuity and creativity can yield many different solutions to problems bounded by a given set of constraints. Collaboration offers us the chance to reap a rich harvest of ideas and solutions germinated in different intellectual soil.

As we at NASA learned during the Apollo program and are re-learning in Constellation, the operation of complex, integrated space systems requires revolutionary thinking in their development and management. Accordingly, we need to develop new manufacturing methods with the ability to operate to a higher, more precise standard of excellence. This *is* rocket science. But it is also art and the industrial capabilities we create as we learn to master this most difficult art ripple throughout our economy. So it is to our mutual benefit to understand how the other spacefaring nations of the world solve the problems posed in the course of mankind's efforts to master spaceflight. We all have much to learn, and we can learn best by doing some of these things together, each of us making our individual contribution, so that all may benefit in direct and indirect ways.

I've lived through this experience. When we initiated the Shuttle-*Mir* program in the early '90s, many of us at NASA felt a bit put out. It was easier to compete with the Russians than to cooperate with them! But we learned over time, and through shared experiences, to trust them to a far greater extent than we had imagined we could. We learned that different doesn't mean bad. We now defer to the space station partners in regard to their design standards, delegated safety panels and remote mission control centers. And we and the

Russians have learned to trust each other enough to alternate space station design reviews and mission commanders with confidence. We're better than we were because of what we have learned that was new to us and "old hat" to our partners.

For these reasons and where we can feasibly promote it, collaboration on the space frontier is the right thing to do, from both an altruistic and a national interest perspective.

That being said, we must recognize certain realities. The United States is firmly committed to ensuring that certain key space and missile technologies—that we possess and others do not—not be used against us or our allies. That priority is higher for us than partnership in space endeavors, a fact that must be understood by all parties involved in any prospective collaboration. I recognize the bluntness of this statement; but I believe that each of us, as spacefaring nations, must respect each other's national priorities and speak openly and honestly with each other if there are differences that hamper our ability to collaborate.

The other major limitation on collaborative programs is the universal constraint of budgetary resources. NASA simply cannot afford everything that our many partners, domestic and international, would like us to do. It is clear to me that partnerships work best when all partners have "skin in the game," each contributing resources toward a common goal that is greater than that which could be easily afforded by any single partner. I believe that such relationships work best when conducted on a "no exchange of funds" basis. For example, NASA is contributing two sensor payloads to India's *Chandrayaan* spacecraft. NASA teamed with the French Space Agency on the Cloud-Aerosol Lidar and Infrared Pathfinder Satellite Observation mission, an Earth science satellite for which we built the laser radar sensors. France integrated the spacecraft and NASA launched it. The reverse will be true for the James Webb Space Telescope; design and integration will be conducted in the United States, but the observatory will be launched on a European Ariane V from French Guiana.

I must admit that this view of partnership is not universally shared. On many occasions since assuming my role as administrator, and especially in connection with our efforts to define and implement Constellation, I have been asked about opportunities for "partnership" when what was really being sought was American investment in the aerospace industries of other nations. To me, partnership cannot be a synonym for "helping NASA to spend its money." We at NASA need partners not subcontractors.

However, there are always exceptions. Soon after my return to NASA in April 2005, I was faced with the choice of continuing to pay the Russian Space Agency for crew and cargo transport to the space station or de-crewing U.S. astronauts. I regarded this (and still do) as an unseemly position for our nation. We are in this position because for the better part of a generation the nation failed to step up to its commitments to build a crew rescue system for the space station astronauts and a replacement for the shuttle. In the words of Admiral Hal Gehman, Chairman of the *Columbia* Accident Investigation Board, "previous attempts to develop a replacement vehicle for the aging shuttle represent a failure of national leadership."

The Russians developed and have operated for many years their *Soyuz* and *Progress* spacecraft. When the shuttle fleet is retired in 2010, there may be no alternative other than to use *Soyuz* for crew transport and rescue. While I do not relish the idea of paying Russia some $900 million in U.S. taxpayer funds through 2011 (and possibly more in later years) the alternative—removing American presence from the space station—is worse. This reliance on Russia, paying them for their increased support of the space station partnership because of America's inability to meet its partnership commitments with American hardware, is one reason why this nation must now invest the time, resources and energy in developing a new U.S. system for crew and cargo transport and why we must bring these systems on line as soon as possible.

If we are to partner effectively in future exploration endeavors, we must establish clear principles for such partnerships. The story above illustrates one of those precepts; to me, it is clear that America cannot partner from the rear. That is not the posture of a great nation.

But however it is done, working together in space helps all of us to realize our common humanity. It shows us that what binds us together is far more important than the issues that separate us.

This certainly can be difficult to keep in mind. Fifty years ago, Americans looked into the sky with fear and trepidation at a small metal orb that was circling our Earth, *Sputnik*. Many Americans felt vulnerable to Soviet missiles, fearing that if the Soviets could place this small satellite in orbit then they could also strike anywhere in the United States. No other adversary had ever produced such a threat to the American homeland and, protected as we were by two oceans, no one in 1957 had ever imagined that anyone ever could. Nikita Krushchev's November 1956 admonition, "We will bury you," reverberated in America's collective consciousness.

Not far from here as he lived in and walked the streets of Georgetown, the junior senator from Massachusetts bore witness to the *Sputnik* crisis 50 years ago. It spurred the creation of NASA and America's space race with the Soviet Union. John F. Kennedy was the first of our nation's leaders to fully appreciate the strategic importance of space exploration. He recognized that the United States trailed the Soviet Union in human spaceflight; and he recognized its significance to the world's perception of America's leadership, saying:

"Those who came before us made certain that this country rode the first waves of the industrial revolution, the first waves of modern invention and the first wave of nuclear power; and this generation does not intend to founder in the backwash of the coming age of space. We mean to be a part of it—we mean to lead it. For the eyes of the world now look into space, to the moon and to the planets beyond, and we have vowed that we shall not see it governed by a hostile flag of conquest but by a banner of freedom and peace. … In short, our leadership in science and in industry, our hopes for peace and security, our obligations to ourselves as well as others, all require us to make this effort; to solve these mysteries; to solve them for the good of all men; and to become the world's leading spacefaring nation."

President Kennedy's insights have stood the test of time; certainly others in the world understand them even as the import of that challenge to our nation has faded in the American collective consciousness. It has been 35 years since Americans Gene Cernan and Harrison Schmitt walked on the moon in December 1972. Thirty-five years. Some young people today actually question whether we ever really set foot on the moon, whether it was all a hoax. Thirty-five years ago, who would have guessed that such a thing could ever have occurred?

I have on many occasions offered the blunt opinion that America made a mistake of strategic importance when, in the early 1970s, we dismantled our nation's technical capability to build the Saturn rocket, Apollo capsules and lunar landers—the means by which NASA met President Kennedy's challenge and defined his lasting legacy. The space shuttle we first flew in 1981 is an amazing machine with unparalleled capability. It is, however, limited to low Earth orbit by its very design. Now our nation must rebuild the capability to journey once again beyond low orbit to see and explore the universe with our own eyes and hands, not just with robotic ones.

I will again quote Hal Gehman in the report of the *Columbia* Accident Investigation Board: "The U.S. civilian space effort has moved forward for more than 30 years without a guiding vision." That was a damning statement highlighting a lack of leadership in space policy reaching to the highest levels of our nation for over a generation. Based on the policy debate that ensued after the *Columbia* accident, President George W. Bush committed our nation to fulfilling our obligations to our international partners by finishing the space station and invited them and others to join the United States in our return to the moon and future ventures to Mars and beyond. Congress codified this direction into law with the NASA Authorization Act of 2005, a copy of which hangs outside my office. In my opinion, this is the best direction NASA has received from Congress in 40 years or more and is a palpable recognition that "space" is a strategic interest of the United States. And last month on the floor of the United States Senate, a large, bipartisan group of senators expressed their strong support for NASA's mission and the challenges we face.

NASA is taking the first steps in this long journey by fulfilling our commitments to our international partners with the space station, retiring the space shuttle and building the new Orion and Ares crew and launch vehicles to support the station and return to the moon. We are also encouraging and spurring a burgeoning commercial space industry in the United States with the space station. Combined, this is the greatest management challenge NASA has ever faced.

However, we are now beginning that quadrennial political season in Washington and some space policy pundits and critics have begun to speculate that we do not have the national will to return to the moon or to venture astronauts beyond low orbit—this time to stay. They argue that NASA's budget, a mere 6/10ths of 1 cent of every federal dollar, is too much. In their minds, Gene Cernan would indeed be the last American to set foot on Earth's moon.

If that future comes to pass, then I will tell you flatly that we will have ceded our leadership on the frontier of space exploration to other countries through softness, complacency and a lack of national will. If that happens then America's best days are indeed behind us.

I believe that talk of retreating again to low Earth orbit merely foments pointless discord, setting aside for the sake of partisan politics the strategic foresight of what is important to our nation. Quite simply, for the United States to be anything other than *the* leader on the space frontier is a mistake of historic proportions. We are a wealthy nation both economically and intellectually. Leadership in space cannot be taken from us; we can only let it slip ineluctably away by failing to recognize its importance to our national security, our technological superiority, our industrial base and our ability to compete favorably on a global scale. If that happens, we won't live to know the cost of it but our children and grandchildren will to their detriment.

I would like to conclude with President Kennedy's advice on November 21, 1963, almost 44 years ago on the day before he was assassinated:

> For more than 3 years I have spoken about the New Frontier. This is not a partisan term, and it is not the exclusive property of Republicans or Democrats. It refers, instead, to this nation's place in history, to the fact that we do stand on the edge of a great new era, filled with both crisis and opportunity, an era to be characterized by achievement and by challenge. It is an era which calls for action and for the best efforts of all those who would test the unknown.

President Kennedy's challenge to NASA and our nation continues today. If we want to be a nation with which other nations will want to collaborate, we must continue to show the bold leadership and commitment to action called for by President Kennedy. The need to take these steps will be seen most clearly if we fail to take them. We can never allow that to happen.

Thank you.

The Next Generation of Engineers

Michael D. Griffin
Administrator
National Aeronautics and Space Administration
Remarks at National Society of Professional Engineers
Professional Development Conference
Washington, District of Columbia

January 19, 2006

The great American Author James Michener, who frequently had interesting observations about our society, once said. "Scientists dream about doing great things. Engineers do them." Although Michener was often right on target, I think this statement requires some modification. Obviously, scientists are capable of doing great things. But perhaps not as obvious is that some of the best engineers in the business are also dreamers, people who refuse to be satisfied with the status quo, who are able to determine new and better ways of achieving grand objectives and then implement them using the disciplines we have all been taught.

We at NASA are going to need these engineers, and plenty of them, to achieve the goals of the Vision for Space Exploration. Two years ago, on January 14, 2004, President Bush committed this nation to a new direction in space and set forth a fresh, clear mission for NASA. The president's directive gave all of us who are privileged to work in this business a challenge bold enough to last a lifetime. Indeed, it is a challenge to last several generations.

NASA is undertaking a program of human and robotic exploration of the moon, Mars and beyond that will enable human beings to do things that have never been done before, see things that have never been seen before and discover things that may never have been dreamed of before. If I were graduating today, I would want to work in the space program for no greater reason than to be a part of these amazing challenges and opportunities.

And in fact, we will recruit the best and brightest engineers out of college to help us develop the next generation spacecraft, launch vehicles and systems that will enable these voyages of exploration to unfold. These exciting missions will motivate today's grade school and high school students to want to work as engineers and scientists in the space program.

Nothing is more important to our future in space. Currently, the engineers of the Apollo era—which ended over 30 years ago—have nearly disappeared at NASA. Our present generation of engineers largely cut their teeth on the space shuttle and space station programs. And even this "baby-boomer" generation of engineers, of which I am one, will for the most part soon be passing from the scene. So we will need the talented young engineers and the promising students coming up through the educational pipeline because the era we are entering must have a steady flow of engineering talent for the next 30 years and beyond.

With this in mind, I challenge the National Society of Professional Engineers to work as an advocate for and a strong partner of NASA in nurturing our schools and universities to establish and maintain excellence in our engineering curricula and motivate a new generation of engineering talent.

While NASA is not the Department of Education, consistent with our charter we do spend nearly $167 million on education initiatives that are targeted to our future workforce needs and to developing the talent, skills and professions necessary to carry out the Vision for Space Exploration. We also spend a like amount in the context of education and public outreach efforts associated with individual space missions. We must ensure that these investments benefit not only NASA but also the aerospace industry as a whole.

To be certain, what we and you do in this regard has broader implications. As the blue ribbon panel of the National Academy of Sciences headed by Norm Augustine has so forcefully pointed out, there has been a steady erosion in investment in the kind of scientific and engineering brainpower that keeps a nation competitive—and a consequent decline in American inventiveness.

So anything that we can do to arrest these trends: to inspire young students to pursue technical careers and to motivate talented foreign-born graduate students to consider staying in the United States and work on the greatest exploration project of the 21st century—would be all to the good.

I'd now like to address how we are organizing our engineering work at NASA to achieve the kind of technical excellence that is necessary to execute our long-term exploration program successfully. As a central organizing principle of our work, and despite the fact that 80 percent of our total funding goes to industry and will continue to do so, I firmly believe that it must be NASA and its engineering staff, not our contractors, who will assume the primary responsibility for making this program work. We are undertaking a multigenerational program of sustained exploration, and we must ask where our intellectual capital should reside. Should it be outside the government in the hands of a prime contractor whose interests may change over the years? Or should it remain in-house, where we can sustain the program's momentum and retain an institutional memory of the system and cost trades that are made and a strong understanding about why the architecture is the way it is? I do not believe that it is wise to contract out these vital functions. Making NASA engineers clearly responsible and accountable for our technical products at the system level will drive our team toward excellence.

Having decided this, we want to provide our engineers with an environment that will help them succeed, by giving them the best possible tools, facilities, training and processes. These will allow them to be competent in assessing risk and assuring mission success.

Already under way is an effort to upgrade our computer assisted design tools and the infrastructure that supports our engineering workforce at the 10 NASA centers. With respect to training, NASA's Academy of Program and Project Engineering Leadership provides a variety of training and learning opportunities for engineers to help develop their competencies and skills throughout the lifecycle of a career in engineering and project management. The curriculum of the Academy employs state-of-the-art methodologies based on the best empirical research and the latest developments in industry. One of the innovations of the Academy is the *Academy Sharing Knowledge* or ASK magazine, which gives NASA managers the opportunity to swiftly tell each other about successes, failures and lessons learned. These "after action reports" were featured in a recent issue of *Government Executive* magazine as a model for what every federal manager should be able to tap into.

But as valuable as this training is, there is no substitute for hands-on experience. Our associate administrator for safety and mission assurance, former astronaut Bryan O'Connor, recently asked a senior Southwest Airlines captain to explain why his airline had the best safety record in the world. The pilot responded by saying, "We always fly manually during the high risk parts of the flight." The pilot then explained that if something goes wrong during takeoff, climb out, approach and landing, he and his colleagues would be better able to react to an emergency situation if they were already flying the aircraft. As a much less accomplished pilot myself, I absolutely believe this observation. And I want NASA to have the same approach. Our engineers will be better able to react to problems in a development activity if they are directly involved. In the past, it has been possible to be an engineer working at NASA over a 25-year career devoted to managing engineering, observing engineering and yet never once doing engineering. That will change. We will give our engineers the opportunity to learn, experiment and succeed or fail on in-house work as part of their normal career progression. This will make them smarter buyers on our contracted efforts and better leaders as they mature.

We must also be able to develop within NASA good processes that will help us execute our mission objectives with careful and sober attention to the management of risk. First and foremost, we will continue to encourage our people to speak up whenever they have safety concerns. And we will listen to and respond to those concerns. The *Columbia* accident and other mishaps have shown that we in this agency have not always listened as carefully as we should have.

The *Columbia* Accident Investigation Board (CAIB) observed that the agency had not been exercising its engineering curiosity sufficiently. Accordingly, we are benchmarking other organizations involved in complex systems engineering projects, such as DOD research, development and engineering projects, and those in the nuclear safety industry. In another response to the CAIB, we are taking a fresh look at some of the hazard analysis and engineering models that had been developed in the past and updating them to incorporate new experience and current thinking.

We are upgrading our ability to provide independent assessments of our work, so that at key steps in a project we can check progress, make appropriate adjustments and catch the things that people miss when they are focused on crucial details but, possibly, are missing the big picture. This gets directly to the issue of programmatic authority versus technical authority in the management of large, complex programs, a topic upon which the CAIB spent considerable time. I'm convinced that it is necessary to have independent technical and programmatic lines of command at NASA because there will always be a healthy tension between the programmatic imperative to accomplish tasks within cost and schedule and the technical imperative to do things perfectly, regardless of cost or schedule. Without this organizational separation, one imperative or the other must dominate always to the detriment of either the project or the institution in my experience. By having this separation, the valid viewpoints of both are preserved to the benefit of both the program and the institution. I believe that this approach will restore our ability at NASA to provide independent technical review of programs in a way that the CAIB found lacking in the shuttle program prior to the loss of *Columbia*.

To provide a further independent engineering assessment capability, we have a group at the Langley Research Center called the NASA Engineering and Safety Center (NESC), comprised of some of the most talented engineers in the agency. This organization draws upon engineers throughout NASA to assist on some of our most difficult technical problems. The NESC has contributed greatly to the shuttle return to flight effort, in part by leading more than a dozen independent assessments of important technical issues. One of these assessments led to the design and building of an insulating wrap, utilizing a sacrificial retainer made of shrink wrap and aerogel, which can be used to preclude the buildup of ice on the space shuttle external tank liquid oxygen feedline bellows. Another assessment, done in conjunction with the Lawrence Livermore National Laboratory, brought to light a potentially catastrophic failure with composite overwrapped pressure vessels due to stress rupture not revealed by earlier studies but unearthed by more recent analyses. The NESC is now working with the shuttle program office on lifetime assessments for the remaining shuttle hardware.

Through diligence like this, we are working to establish standards of technical excellence that will enable a program of the complexity and promise of deep space exploration to move forward over a period of decades. As I said in the beginning, all human knowledge and skill will be needed to push the frontier out beyond low Earth orbit. But as Michener suggested, engineers will have a special role in the doing of it. That's why I am devoting so much of my own time to working with our team to make sure that our engineering workforce is given the best possible opportunity to acquire and demonstrate technical excellence in all of its facets.

In closing, I thank all of you for your commitment to excellence in engineering and for your strong interest in what we are trying to accomplish at NASA.

System Engineering and the "Two Cultures" of Engineering

Michael D. Griffin
Administrator
National Aeronautics and Space Administration
Boeing Lecture Purdue University
March 28, 2007

Most of you will have heard of Baron Charles Percy (C. P.) Snow and will know of his observations on the breakdown in communication between the humanities and the sciences. Trained as a scientist, Snow served as minister for technology under Prime Minister Harold Wilson yet was more famous as an author, with 16 novels and eight works of non-fiction to his credit. He would be near the top of nearly any list of scientifically literate authors or of literarily-talented scientists. Snow developed his theme in *The Two Cultures and the Scientific Revolution* in 1959 and explored it further in *The Two Cultures and a Second Look* in 1963. He decried the decline in standards of higher education and in particular what he viewed as the almost willful ignorance by the modern cultural elite of scientific fundamentals. In a summary of his theme, Snow noted,

> A good many times I have been present at gatherings of people who, by the standards of the traditional culture, are thought highly educated and who have with considerable gusto been expressing their incredulity at the illiteracy of scientists. Once or twice I have been provoked and have asked the company how many of them could describe the, Second Law of Thermodynamics, the law of entropy. The response was cold: it was also negative. Yet I was asking something which is about the scientific equivalent of: 'Have you read a work of Shakespeare's?'
>
> I now believe that if I had asked an even simpler question—such as, What do you mean by mass, or acceleration, which is the scientific

> equivalent of saying, 'Can you read?'—not more than one in ten of the highly educated would have felt that I was speaking the same language. So the great edifice of modern physics goes up, and the majority of the cleverest people in the western world have about as much insight into it as their Neolithic ancestors would have had.

While Snow's criticisms did not go unanswered—most famously by Literary Critic F.R. Leavis—the essential truth of his observations was, and is, widely acknowledged. His elucidation of the "two cultures" has become a societal paradigm, a bumper-sticker phrase to describe the basic cultural separation between the arts and the sciences that is clearly visible to most of us. Even those who know nothing else of Snow's work are probably familiar with this one phrase.

Today, I want to discuss the two cultures that, if we think about it, we find embedded in the profession we call "engineering," and how we are linking them, and must link them, through the discipline known as "system engineering," a product of the American aerospace sector.

Let us first explore the nature of the "two cultures" in engineering. I have always loved the view of the engineering profession captured by the great Theodore von Karman when he said, "Scientists study the world as it is; engineers create the world that has never been." Less eloquently, engineers are designers; they synthesize knowledge to produce new artifacts. Von Karman speaks to what most of us, and certainly most laymen, would consider the essence of engineering: engineers create things to solve problems.

But all of us who are engineers know that the engineering profession also has a rich scientific side, the analysis of these artifacts and the prediction of their behavior under various environmental and operational conditions. Adapting von Karman's observations, it may be said that engineering science is the study of that part of the world created by man.

Sadly, many students have been led to believe that engineering science is engineering! In a curriculum of 120 or more credits leading to a bachelor's degree in a branch of engineering, the typical student is required to take one or maybe two courses in design. Everything else, aside from general education requirements, focuses on the analysis rather than the creation of engineered objects. Graduate education often has no design orientation at all. So, engineering as taught really deals with only a part of engineering as it is practiced.

This trait is so pronounced that engineers who have spent their careers—even widely recognized careers—in design and development, focusing on the creation of objects rather than the creation of papers for publication in refereed journals, are essentially unemployable (hence unemployed) in academia. No matter how well credentialed a practicing engineer may be when the inevitable search committee meets to rank the applicants for a department chair or a tenured position, it is a rare designer who can offer even the minimum of "academic" qualifications expected of an applicant for the position of assistant professor.

Some universities have recognized this inherent bias and its consequences for the training of their students and have sought to remedy it by creating titles such as "professor of practice" or similar appellations. But it is a truism that the longer the title, the less important the job. So this term serves only to emphasize the point that these particular faculty members are not "real" professors hired and promoted on their merits in a straight-up competition among all candidates. One wonders if this is the message we really want to send to those who will design (or not) the world of the next generation.

But if the present excessive focus on engineering science in the engineering curriculum is of concern, it is nonetheless true that the fundamental difference between modern engineering and that practiced prior to the Enlightenment is the development of formal analytical methods and their application to manmade objects. This has allowed the prediction of performance, and the limits of that performance, in the environment in which a given device must function. It has allowed the refinement of designs through methods more sophisticated than the trial-and-error techniques to which our ancestors were limited. It has enormously shortened the time required for a design cycle for the objects we create. A control system engineer might say that the formal methods of engineering science have produced an enormously improved feedback path for the engineering design loop. More simply, engineering science has taken engineering beyond artisanship.

But, interestingly, the development of formal methods has not altered in any way the fundamental nature of design, which still depends, as it did in antiquity, upon the generation of a concept for a process, technique or device by which a given problem might be solved. The engineering sciences have provided better, and certainly quicker, insight for the designer into the suitability of the concept than can be provided solely by building it and examining its performance in its intended application. But a human being must still intuit the concept. We have no idea how we do that. And until we do, we have little hope of developing a formal method by which it can be accomplished.

It must be said that some progress in this area has been through research into "genetic algorithms" that use the tools of engineering science and mathematical simulation to explore the consequences of iterative random changes to a given design. The performance of the design is evaluated against objective criteria. If a change results in a net improvement, it is retained; otherwise, it is discarded. In this manner, the design evolves to a higher state of suitability to its intended environment through the pressures of artificial, rather than natural, selection. Modern engineering analysis tools offer the ability to conduct what is essentially a very large number of randomized design cycles in an acceptable period of time.

But this process does not seem, at least to me, to be much akin to the intuitive synthesis of a human brain when it leaps almost instantly from a perception of a problem to an idea for its solution. Creativity, used in this sense, remains thus far the sole province of biological computers.

However, my colleague, NASA Associate Administrator Lisa Porter, has pointed out to me that precisely because genetic algorithms work differently and produce different results than would a human designer, they can offer new, unusual and potentially useful solutions for consideration by humans. So as the field of genetic algorithms matures, it may well be that the methods of engineering science will yield solid contributions to the synthetic aspect of engineering.

But at least for now, there remains an artistic side of engineering; and it is fully as much an art for its practitioners as any painting, sculpture, poem, song, dance, movie, play, culinary masterpiece or literary work. The difference between the cultural and engineering arts lies not so much in the manner of creation of a given work but in the standards by which that work is judged. In the humanistic disciplines, human aesthetics sets the standard by which merit is assigned to a finished product. In the end, aesthetic sensibilities vary with place and time and are ultimately matters of opinion. The role of opinion in evaluating a work of engineering is, by comparison, much restricted. In engineering, more objective methods are employed to judge the degree to which the completed work meets the standards established for it or fails to do so.

This brings us to the role of failure in engineering design. Regardless of the sophistication of the analytical methods brought to bear, they are applied to a theoretical model of a device operating in a theoretical model of the real world. The model is not reality, and the differences produce opportunities for the real device to fail to operate as intended in the real environment. An evolutionary biologist might say that the gap between model and reality is an environmental niche in which failure, like a new species, can thrive.

Civil Engineer and Author Henry Petroski has, in a series of essays and books, explicitly noted the crucial role of failure in producing ultimately successful designs. In *Success Through Failure: The Paradox of Design* and other works, Petroski establishes the point that new designs or successive iterations and refinements of a basic design have as their essential purpose the elimination of failure modes known to be inherent in earlier designs. He further argues, by means of many examples, that designers must go beyond merely ensuring success; they must strive to anticipate the ways in which a design might fail. Great designers and successful designs incorporate, in advance, methods to mitigate such anticipated failures.

But in recent decades human artifacts have become increasingly complex, building upon and extending former art and especially combining disparate elements of established art in new ways. This has been accomplished at an astonishing pace, a cause and a result of Moore's Law, the approximate 2-year doubling time of computational throughput, which has held sway for several decades. While a large bridge cannot properly be considered a "simple" structure, involving as it does the interaction of thousands of component parts, it clearly pales in complexity relative to, say, a space shuttle, which relies for its success upon the interaction of millions of parts derived from a dozen technical disciplines.

Failure in complex systems can arise in so many more ways than in simpler systems that the quantitative difference ultimately produces qualitatively different behavior. It becomes unreasonable to expect, other than through the harshest of hindsight, that a particular failure mode might have been or ought to have been anticipated. Indeed, results from the modern study of complexity theory indicate that complex systems can experience highly non-linear departures from normal state-space trajectories—for example, "failure"—without anything being wrong.

Among the first to study complex engineering systems was Charles Perrow, in the landmark work *Normal Accidents*. Perrow argued that adding additional processes, safety measures and alerts to complex systems—the traditional design approach to improving system safety—were inherently flawed because for complex, tightly coupled systems and organizations, failure is inevitable.

Perrow is a sociologist, not an engineer, but his points are well taken. Those of us who are aviators, or who are familiar with the history of aviation, can point to numerous high-profile accidents where the crew became occupied with minor anomalies and their warning systems, only to fly a perfectly good airplane into the ground. Most of us can also cite analogous incidents from other fields.

Yet, we have evolved complex systems for good reasons, and we will clearly continue to do so. The modern air transport aircraft is an incredibly complex device, and the system within which such aircraft operate is far more so. But in the last five decades this system has revolutionized world society, culture and economics. It will not be shut down merely because it cannot be made perfectly reliable. Nor will we do so with any of the other complex appurtenances of modern society that did not exist a century ago but are now deemed essential. So, if we are not to eschew the use of complex systems, how do we make them as reliable as possible?

I believe that the answer to the above question is "system engineering." This is an entirely appropriate answer for the Boeing Lecture here at Purdue University, for system engineering has evolved as a discipline of modern engineering from its roots in the American aerospace system development culture.

System engineering and its allied discipline of systems management are treated from an historical perspective in the excellent text by Stephen Johnson, *The Secret of Apollo*. Johnson retraces Petroski's path, showing the development of system-oriented disciplines to be the natural reaction to the failure of early, complex aerospace systems, including large aircraft, ballistic missiles, and spacecraft.

From its first introduction into the engineering lexicon, "system engineering" has been a question-begging term. In earlier times, it was considered by many in the traditional engineering disciplines to be a category without a subject matter. Even today I find the term to be, in my opinion, misused and misunderstood by many who claim to be practitioners of the art. So, having spent what I believe to be the most productive part of my career as a system engineer, let me say a few words about what I believe system engineering is, and what it is not.

System engineering is the art and science of developing an operable system capable of meeting requirements within imposed constraints. The definition is somewhat independent of scale, and so these words are useful only if one understands that it is the big-picture view, which is taken here. We are talking here about developing an airplane, a spacecraft, a power plant and a computer network. We are not talking about designing a beam to carry a particular load across a known span.

System engineering is a holistic, integrative discipline, wherein the contributions of structural engineers, electrical engineers, mechanism designers, power engineers and many, many more disciplines are weighted and considered and balanced, one against another, to produce a coherent whole that is not dominated by the view from the perspective of a single discipline. System engineering is about tradeoffs and compromises, about generalists rather than specialists.

System engineering is not about the details of requirements and interfaces between and among subsystems. Such details are important, of course, in the same way that accurate accounting is important to the chief financial officer of an organization. But accurate accounting will not distinguish between a good financial plan and a bad one nor help to make a bad one better. Accurate control of interfaces and requirements is necessary to good system engineering, but no amount of care in such matters can make a poor design concept better. System engineering is about getting the right design.

Complex systems usually come to grief, when they do, not because they fail to accomplish their nominal purpose. While exceptions certainly exist, it remains true that almost all systems which proceed past the preliminary design phase will, in fact, accomplish the tasks for which they were explicitly designed. Complex systems typically fail because of the unintended consequences of their design, the things they do that were not intended to be done. The Second Law of Thermodynamics is sufficient to guarantee that most of these things will be harmful! I like to think of system engineering as being fundamentally concerned with minimizing, in a complex artifact, unintended interactions between elements desired to be separate. Essentially, this addresses Perrow's concerns about tightly coupled systems. System engineering seeks to assure that elements of a complex artifact are coupled only as intended.

C.P. Snow believed that mutual comprehension and appreciation between the arts and the sciences, which had existed in earlier times, had been erased by his time. He did not find a means to restore it. I sometimes think that the gap between synthesis and analysis in engineering is as wide as that between the arts and the sciences of Snow's "two cultures." But the fact remains that designers simply do not think or work in the same way as analysts and this does on occasion produce a certain cognitive dissonance. When it occurs in the context of a complex system development, catastrophe is a likely result.

System engineering is the link, which has evolved between the art and science of engineering. The system engineer designs little or nothing of the finished product; rather, he seeks a balanced design in the face of opposing interests and interlocking constraints. The system engineer is not an analyst; rather, he focuses analytical resources upon those assessments deemed to be particularly important from among the universe of possible analyses which might be performed but whose completion would not necessarily best inform the final design. There is an art to knowing where to probe and what to pass by, and every system engineer knows it.

Like other branches of engineering, system engineering has evolved out of the need to obviate dramatic failures in complex systems. Such failures are not new. One of my favorite books is a fascinating text entitled, *Structures: or, Why Things Don't Fall Down*, by Professor J. E. Gordon of the University of Reading, England, written in 1978 at the end of Professor Gordon's long career as a structural analyst. It is aimed at a level appropriate to an intelligent technical professional in any field. I recommend it highly. Regarding the matter of spectacular engineering failures, I quote Professor Gordon (pp. 352–53):

> There are, of course, a certain number of great dramatic accidents which, for a while, monopolize the headlines. Of such a kind were … [numerous disasters follow] … These are very often intensely human and intensely political affairs, caused basically by ambition and pride. … One can at once recognize a certain inevitability about the whole procedure. Under the pressure of pride and jealousy and ambition and political rivalry, attention is concentrated on the day-to-day details. The broad judgements, *the generalship of engineering*, [my emphasis] end by being impossible. The whole thing becomes unstoppable and slides to disaster before one's eyes. …

In 36 years of engineering practice, of many kinds and in many situations, I have not seen a more appropriate assessment of what is truly important in engineering. We must of course get the details right. However, to be a complete engineer, one must also master what Professor Gordon calls "the generalship of engineering."

I will be frank. Educators, and I include myself for I have spent many years as an adjunct professor at various institutions, are far less certain how to teach “generalship” than we are of how to teach the laws of thermodynamics. And yet it is clear that an understanding of the broad issues, the big picture, is so much more influential in determining the ultimate success or failure of an enterprise than is the mastery of any given technical detail. The understanding of the organizational and technical interactions in our systems, emphatically including the human beings who are a part of them, is the present-day frontier of both engineering education and practice.

Generalship of Engineering

Michael D. Griffin, Administrator
National Aeronautics and Space Administration
Commencement Address
Johns Hopkins University
Whiting School of Engineering

Baltimore, Maryland

May 24, 2006

Thank you for inviting me to speak here at the Whiting School of the Johns Hopkins University. I am twice an alumnus of this university, with my first Hopkins degree granted before most of you who are here today were born. So it is always a pleasure to be back here at my alma mater. I hope you will be able to look back on your years at the university with as much affection as I do on mine.

In that vein, I'd like to take a moment to applaud your success. You have earned a degree at one of our nation's finest academic institutions, and you have every right to be proud of that accomplishment. For the rest of your life, you will look back upon your time here, and increasingly you will come to understand how these years have helped to shape the course of the rest of your life. Graduation, at any level, is a huge milestone.

And I believe that you should be especially proud to be graduating as engineers. With all due respect to other fields of endeavor, I will bring to your attention one of Robert Kennedy's favorite quotes: "Some men see things as they are, and ask 'why'? I see things that never were, and ask 'why not'?" Though not of a technical origin, no phrase more perfectly captures for me the role of the engineer in human society. As Theodore von Karman put it, "Scientists study what is; engineers study that which has never been." Engineers are artists, with the world of science, mathematics and technology as their easel, brush and canvas.

But graduation as an engineer is only one of many milestones to come, and with this honor comes responsibility. One of those responsibilities is to step up to the leadership opportunities that will come your way. So, how do we abstract from our experiences as students the lessons that will be most helpful to us 10, 20 or 30 years in the future?

In addressing this group, as always on similar occasions, I am faced with a key question: What can I talk about in 20 minutes or so that might be of some value to you, something that you might recall in 20 years? It's a difficult problem, if you think about it. And I don't believe the answer lies in a further exposition of Shannon's Theorem, Bernoulli's Principle, Hooke's Law, or some other detail taken from your engineering study. Rather, I think the answer more properly lies in the other direction, in an examination of our profession from the broadest, rather than the narrowest, possible viewpoint.

If we are observant, we realize that engineering and the engineering profession are often in the news, and more often than not in unfortunate ways. For example, in my own field, the resumption of space shuttle flights with STS-121 next month will be big news as it was with STS-114 last July. But the biggest news about last year's flight was the unexpected loss, yet again, of a sizeable chunk of foam from the shuttle's external tank. Similarly, the successful return of the robotic Stardust probe, with samples of matter from interstellar space, made news. But the prior year's 200 mile per hour impact of the Genesis spacecraft in the Utah desert, when its parachute failed to open made headlines. The plain fact is that engineers and their creations are, in today's world, simply expected to be successful, to perform without fail.

The good news is that in the vast majority of instances, we have indeed learned to make it so. The bad news is that, when we fail, it is often in a very serious and public way. And the worst news of all is that, most often, our greatest failures are rooted not in our technical knowledge but in our humanity.

So often, we fail greatly for the most humiliating reasons. The Hubble Space Telescope was placed in orbit in 1989 at enormous expense, the product of a career for many of its creators. And it arrived with its vision brilliantly flawed, its optics perfectly formed to exactly the wrong specifications.

More personally, I will note that we in the space community lost seven friends and colleagues on Space Shuttle *Columbia*, just over 3 years ago. This loss was made all the more difficult to bear by the knowledge, so starkly outlined in the report of the *Columbia* Accident Investigation Board, that *Columbia* was brought down by many of the same human flaws as was her sister ship, *Challenger*, in 1986, in one of the classic failures of engineering management.

There are many—too many—other such examples. As our technical capabilities increase seemingly without bound, it becomes ever more obvious that the next great barrier we engineers must overcome is of our own, rather than nature's, making. As the bard said, "The fault, dear Brutus, lies not in our stars, but in ourselves, that we are men. ..."

What causes these things? Are they failures of engineering or of management? And does it matter?

Dramatic engineering failures are of course not new. One of the best books I have read is a fascinating text entitled *Structures: or, Why Things Don't Fall Down*, by Professor J. E. Gordon of the University of Reading, England, and written in 1978, at the end of Professor Gordon's long career as a structural analyst. It is aimed at a level appropriate to an intelligent technical professional in any field. I could not recommend it more highly. Regarding the matter of spectacular engineering failures, I will now quote Professor Gordon (pp. 352–53):

> There are, of course, a certain number of great dramatic accidents which, for a while, monopolize the headlines. Of such a kind were the Tay Bridge collapse in 1879 ... [other disasters follow]. ...

> These are very often intensely human and intensely political affairs, caused basically by ambition and pride. … One can at once recognize a certain … inevitability about the whole procedure. Under the pressure of pride and jealousy and ambition and political rivalry, attention is concentrated on the day-to-day details. The broad judgements, *the generalship of engineering*, [my emphasis] end by being impossible. The whole thing becomes unstoppable and slides to disaster before one's eyes. ... People do not become immune from the classical or theological human weaknesses merely because they are operating in a technical situation. …

In 35 years of engineering practice, of many kinds and in many situations, I have not seen a more appropriate assessment of what is truly important for each of you to bear firmly in mind throughout your engineering career. I do not say that you should not work, and work hard, to get the details right. You should. But these matters you have been taught. You know how to care for them. However, to be a complete engineer, you must also master what Professor Gordon calls "the generalship of engineering."

I will be frank. We as educators, and I include myself for I have spent many years as an adjunct professor at various institutions including this one, are far less certain how to teach you this "generalship" than we are of how to teach you, for example, the laws of thermodynamics. And yet it is clear that an understanding of the broad issues, the big picture, is so much more influential in determining the ultimate success or failure of an enterprise than is the mastery of any given technical detail. The understanding of the organizational and technical interactions in our systems, emphatically including the human beings who are a part of them, is the unconquered frontier of both engineering education and practice.

I will leave you with another of my favorite sayings: "Managers do things right. Leaders do the right things." Here at the university, you have learned to do things right. Now you must go forth and apply this knowledge; and while you are about it, you must learn also to do the right things. I wish you all good fortune in this quest.

NASA and Engineering Integrity

Michael D. Griffin
Administrator
National Aeronautics and Space Administration
Wernher von Braun Memorial Symposium
American Astronautical Society
October 21, 2008

While some people like to think that Washington, D.C is the center of the universe, any aerospace engineer knows that it's more fun, more immediately rewarding, to be where the action is, to be part of a great team where great things are being built, contributing to a great cause that you can see in front of you. Yes, Washington has a crucial role to play in the management of any federal agency, certainly including NASA, but it is not where the action is, especially when it comes to building, or in this case rebuilding, the capability for our nation to propel Americans beyond low Earth orbit back out to the New Frontier of which President John F. Kennedy spoke.

No collection of books contains all the knowledge one must have to succeed in spaceflight. The unwritten lore of space system design and engineering fills volumes, all stored in irreplaceable human minds. And that is just the technical stuff. Engineering texts do not touch the most important of all elements in the success or failure of any space mission, the human system.

It is people who power our spacecraft, who build the machines to carry out every complex space mission. It is people that matter—how we organize and utilize their energy, how we bring their skills to bear, how we listen to and work with each other and how we inculcate an ethos where the best ideas take flight. It is people who have created the art and science of space vehicle design, the most challenging engineering problem of our age.

Now, while I know how important the management of the human system is to the success of any endeavor, I will not pretend to understand it very well, even in the relatively narrow context of aerospace program management. If I did, I could not possibly cover the topic in a single speech. Tomes have been written on effective management, most of which omit entirely the more crucial, yet even more nebulous, quality we call "leadership." These terms are not synonymous. To me, management is "doing things right" while leadership is "doing the right things." But they share a common element that I believe to be the foundation of effective human organizations: integrity. And that is what I would like to talk about today.

I am, of course, speaking about ethical decision making in our professional lives, about creating a culture within which all can act and speak with openness and honesty, about embracing the responsibility for our statements and actions. Integrity matters enormously. I personally believe that without it, there is nothing else which does matter.

Long stated as one of the core values of our agency, it is nonetheless hard to define integrity in the abstract. It is much easier to recognize it when we see it. It is a quality not well suited to self-assessment, a quality for which we are more easily judged by others than by ourselves. I'm sure that each of you has observed acts of notable integrity as well as cases where people fell well short of expectations. We should examine the differences, make note of what integrity "looks like" in practice and strive for it.

In engineering practice, integrity is speaking up in a meeting when you do not believe the facts match the conclusions being reached or that certain facts are being ignored. Integrity is following the data. Integrity is refusing to fall in love with your own analysis, admitting that you are wrong when presented with new data that should alter your earlier view. Integrity is keeping a promise or commitment or, when circumstances change, explaining why an agreement cannot be kept. Integrity is walking into your boss's office, closing the door and speaking with frankness, openness and honesty—and listening the same way.

Integrity is being willing to put your badge on the boss's desk when you believe that an ethical breach warrants such drastic action.

Integrity is the foundation upon which other human virtues are built—trust, credibility, leadership—and that foundation can be damaged for a very long time, even irreparably, with the slightest crack in a person's or an organization's integrity. As a case in point, my long-time friend Arnie Aldrich, manager of the Space Shuttle Program at the time of the *Challenger* accident, has recently written about some of the long-term consequences of that event:

> In addition to the tragic and unforgettable loss of life, the *Challenger* accident had, and continues to have, momentous effects on United States and international space programs. In the near term, it led directly to an unprecedented restructuring of the Space Station *Freedom* program resulting in extensive redirection, massive delays and huge cost overruns. This was half a dozen years prior to the second massive restructuring of the space station under NASA Administrator Daniel Goldin, which likely never would have occurred. A strong case can be made that if the *Challenger* accident had not occurred the space station would have flown and become operational a decade earlier than what has transpired, with attendant cost savings and opportunities to expeditiously move forward with future plans and programs. Also, the shuttle was reined in from its full potential with decisions to move away from the Department of Defense (DOD) and commercial customers, a large, flexible onboard upper stage and west coast launch capability. In the longer term, the play out of these events continues today as NASA struggles to plan for effective space station operations without the space shuttle while attempting to move forward with a vast new program of human space exploration beyond Earth orbit. The *Challenger* accident changed the course of history and

the nature of both national and international space programs even as these programs continue to evolve in the 21st century. The full impact of the *Challenger* launch decision is still unfolding.

And, again, in late August 2003, when the *Columbia* Accident Investigation Board's report discussed NASA's organizational flaws, they noted that:

> The organizational causes of this accident are rooted in the Space Shuttle Program's history and culture, including the original compromises that were required to gain approval for the shuttle program, subsequent years of resource constraints, fluctuating priorities, schedule pressures, mischaracterizations of the shuttle as operational rather than developmental and lack of an agreed national vision. Cultural traits and organizational practices detrimental to safety and reliability were allowed to develop, including: reliance on past success as a substitute for sound engineering practices (such as testing to understand why systems were not performing in accordance with requirements/specifications); organizational barriers which prevented effective communication of critical safety information and stifled professional differences of opinion; lack of integrated management across program elements; and the evolution of an informal chain of command and decision-making processes that operated outside the organization's rules.

Sadly, this damning indictment of NASA's engineering management culture was supported by the facts of the accident. It was in part for this reason that we implemented a new governance model within NASA. The approach that we chose is modeled on that used in the Office of Space Flight by former NASA Associate Administrator George Mueller during the Apollo era. It places our mission directors and center directors on par with one another and allows for the elevation of concerns and dissent by means of at least two pathways through equally-empowered programmatic and institutional lines of authority.

This was a necessary but not sufficient condition to obtain the "independent technical authority" demanded by the *Columbia* Accident Investigation Board. Remembering that it is the people who count, I then worked—and continue to work—to select individuals of high integrity and established technical acumen *in spaceflight* to fill senior leadership positions. Too often in the past, senior managers at NASA have been allowed to begin their careers in the space business at the top. It doesn't work. An effective organization must be led by people who offer both unquestioned integrity *and* relevant domain expertise. If I were to be granted only one "legacy" for my tenure as NASA administrator, I would want it to be this.

That brings me to certain accusations, both overt and subtle, that I have seen recently in various media, questioning our approach in developing the Constellation launch vehicle architecture with its shuttle-derived Ares I and Ares V. The *Orlando Sentinel*, in an article on June 22, accused NASA of "trying to stifle dissent about alternatives" to the Ares design. That's certainly "overt." More subtly, my friend and colleague Lon Rains, Editor of *Space News*, published on July 21 an editorial entitled "No More Studies," a premise with which I certainly agree. But buried within Lon's editorial are a few words that are actually more troubling to me than those offered by the *Sentinel* precisely because I know that Lon was, on this occasion, actually attempting to speak on our behalf. Lon wrote:

> TeamVision Chief Executive Stephen Metschan naturally questions the objectivity of NASA's analysis and is calling for an independent review of Direct 2.0 versus the Ares architecture, with the aim of convincing the next Congress and presidential administration to set NASA on a new course as early as next year.
>
> It is perfectly understandable that Mr. Metschan would be suspicious that NASA's analysis was unfairly skewed in favor of Ares. …

Oh, really? Why is that, exactly? Why is it "natural" to question NASA's objectivity? Why is it "perfectly understandable" for someone to be "suspicious" that analysis by NASA was "unfairly skewed?" This is the same as saying that it is to be presumed that NASA does not act with integrity—and I know that Lon did not intend to imply such. But is that what some people really believe?

Since my thesis in this speech is that nothing matters more than personal and organizational integrity, let's take a moment to review the bidding from the top down. NASA is a federal agency, an arm of the United States government. NASA employees don't get stock options; they don't get bonuses for concluding mergers and acquisitions; and they do not have financial interests in the industrial concerns that actually implement about 85 percent of the work managed by the agency. We go to tremendous lengths to ensure that NASA employees do not have real or perceived conflicts of interest in connection with their work assignments, to the point that employees must fully disclose their financial investments, regularly attend ethics training and sign legally binding oaths attesting to the absence of conflicts of interest.

What NASA employees do have, to varying degrees, is executive power delegated by Article 2 of the Constitution and specified in great detail through many laws. NASA is *the* entity charged with the implementation and management of government civil space development activities. Public funds allocated to meeting U.S. civil space policy objectives are spent largely according to the technical and programmatic judgment of NASA civil servants as to how it can best be done. NASA employees have the power to decide such issues.

That, of course, is what some critics, many of whom who *do* have financial interests in the outcome of the decisions we make, find to be objectionable. But the management of appropriated funds to accomplish national policy objectives is the very purpose of executive branch agencies. Making decisions in connection with such matters is a core function of government; and for civil space programs, that function is performed by NASA. If we didn't have a

NASA, we'd have to invent one or assign the required functions to some other government entity. The key feature to which some apparently object—that decisions about the allocation of public funds to some alternatives in preference to others are made by government employees—would remain. Only the names would change.

Now, I am not so naïve as to believe that NASA managers are exempt from Lord Acton's observation that, "Power tends to corrupt and absolute power corrupts absolutely." It is absolutely necessary to have interlocking checks and balances between the executive and legislative branches of government as specifically provided by the framers of the Constitution. Our Constitution provides that NASA as an executive branch agency should be overseen by the people's elected representatives. It is. Quite thoroughly, I might add.

But with that said, it is trivial to observe that there can never be enough oversight, never enough checks and balances, never enough watchers, to restrain a large group of people who are determined to behave badly. The effectiveness of institutions generally (and certainly of government institutions) is heavily predicated on a fundamental tenet: most of the people, most of the time, are trying to do their jobs well and fairly. The assumption of well-intentioned competence in government must be the norm, not the exception, in the functioning of a democratic society.

We at NASA cannot possibly make everyone happy with our decisions. Most decisions will produce an unhappy outcome for someone. However, that unhappiness is not by itself a symptom of incompetence, bad intentions or a lack of integrity on our part. Allocation of public funds to any particular alternative inevitably leaves aggrieved parties who believe, with their own logic and passion, that their proposed alternative was the superior choice. It is not reasonable to expect that responsible managers can make decisions pleasing to all interested parties. What the taxpaying public and its elected representatives (our overseers) can and do expect from NASA can be summarized in two words: objective expertise.

NASA cannot be effective as an organization if the decisions of its managers are judged by the space community to be generally lacking in either competence or fairness and that is why such criticisms in *Space News*, *The Orlando Sentinel* and elsewhere, especially the blogosphere, are deeply disconcerting. If it is not obvious that objective expertise underlies NASA decisions and actions, then the civil space program will grind to a halt in response to one searching examination after another by various other governmental entities that claim the right of agency oversight and can make it stick. Thus, it is incumbent upon us to be able to explain how a decision was reached, why a particular technical approach was chosen or why a contract was awarded to one bidder instead of another.

We have all lived through times, some of them recent, when technical competence at NASA was called into question. But today I believe that is not an issue. The management team in place at the agency today is, I believe, second to none in our history. And I think that most of those with even more gray hair than I, who have worked with NASA over the decades, share this belief.

That then leaves the question of objectivity, which of course is exactly the point of comments about "stifling dissent" or "unfairly skewed" analysis. Such accusations are deeply troubling because, in the end, they are accusations that we lack integrity. They chip away at the foundation of the high-integrity organization we strive to build at NASA. The efficacy of our team is predicated upon our ability to "follow the data," to communicate constructively the differences of technical opinion throughout the organization. Accusations to the contrary such as those in the mainstream media or as found on many Web postings reverberate as echoes of lessons *not* learned from the *Challenger* and *Columbia* tragedies.

Because these tragedies are still fresh in our collective consciousness, nothing better serves to cause attention to be focused on NASA's choices—a fact not lost upon those who object to our choices. Such accusations are strong claims indeed. They require strong justification by the accuser and a clear response from the accused. As a manager, I need facts when such charges are levied. Otherwise, it is impossible for me to address them, to prove or disprove their validity or to provide a cure for the cause if there is one.

What must be understood is that differences of technical opinion based on a given set of facts are common among engineers. Such differences of opinion do not mean that the data is "unfairly skewed." A decision by a manager to follow one path rather than another is not evidence of "stifling dissent." To do our jobs, to make forward progress, we must make decisions every day on matters that, unlike the problems in most textbooks, do not have clear, simple, right or wrong answers found in the back of the book. Judgment calls are required; we then often wait years to find out whether they were correct. Not everyone has a taste for the kind of pressure that this brings to bear on senior institutional and program managers but it is inherent to the nature of our business.

Allow me to offer a specific example of how false accusations can be made by taking selective snippets of information out of context. Managers of large, complex projects such as the Ares rocket development use simple "stoplight charts" with red, yellow and green as useful indicators as to where management attention might best be focused. That's all we use them for. They do not begin to convey the subtleties and complexities of managing technical and programmatic risks. But such charts taken out of engineering and management context from internal NASA briefings are regularly featured on various blogs generally accompanied by uninformed and typically anonymous judgments that the Ares rocket will never work and by accusations of lying and malfeasance by NASA managers. Of course, no supporting evidence is ever offered.

So, differences of engineering opinion are cited as evidence of lying, of malfeasance? This is not how any of us were taught to conduct an engineering discussion. Quite frankly, it is demeaning to the profession.

I wonder what Webb or Seamans, von Braun or Gilruth, Mueller or Low or Kraft would have thought if they had had to deal with such vitriol during Apollo? Viewed in hindsight, the success of Apollo can appear to be an unbroken record of progress from President Kennedy's speech to Neil Armstrong's first footstep. But it was hardly so. It took those folks—heroes in our business—18 full months after Kennedy's declaration of the lunar goal merely to determine that lunar orbit rendezvous would be the best flight mode. The original Apollo spacecraft design, with its embedded assumption that Earth orbit rendezvous would be used, had to be substantially modified. An unanticipated procurement of a completely new vehicle, the lunar module, had to be conducted some 2 years after the Apollo program was supposedly well underway. Combustion instability plagued the F-1 engine well into its development and pogo oscillations nearly destroyed the Saturn V on its second mission—an event, by the way, that resourceful flight controllers managed to turn into a success anyway by making great decisions literally "on the fly." But the managers and engineers of that era pressed on, solved the Saturn V's technical problems, and sent three men around the moon on its *very next flight*. If there had been blogs in the 1960s, they would have had so much grist for their mills that they wouldn't have known where to start.

And by the way—just in case anyone has forgotten—Apollo actually turned out pretty well.

So let's fast-forward to the present. Our choices for the shuttle-derived Constellation launch architecture have been especially subject to external criticism by those who would have preferred a different outcome. I strive to be objective in considering the data before me so let's look at the data we used to make the decisions we made.

The probabilistic risk assessment for the shuttle-derived Ares I Crew

Launch Vehicle showed it to be twice as safe as an Evolved Expendable Launch Vehicle (EELV)-derived system for missions to the International Space Station and the moon. The analysis for the shuttle-derived Ares V Heavy-Lift Launch Vehicle showed it to be approximately 1.4 times more reliable than any EELV-derived concept we saw.

If we were to try to undertake a lunar mission using existing EELV systems, at least seven launches would be required to conduct one lunar mission and more than 30 would be required to mount a future Mars expedition. That is not a realistic concept of operations.

If we were to extend existing EELVs to meet our requirements, the development cost would be higher than with the shuttle-derived approach. While a new upper stage would be needed in either case, the Atlas V was preferred over the Delta IV due to its more straightforward development path but at a cost 25 percent higher than the shuttle-derived approach. We would need changes such as pad modifications for crew access, booster structural modifications, improved flight termination and integrated health management systems and a new flight dynamics database to deal with the new outer mold line and abort scenarios. We would need to invest in facilities for U.S. co-production of the Russian RD-180 engine or accept long-term dependence upon Russia for a critical capability by continuing to purchase it directly. The latter course further implies the receipt of a perpetual waiver of Iran, North Korea, Syria Nonproliferation Act (INKSNA) legislation from Congress for such purchases. If we were to make the necessary changes to the EELV, the new vehicle would differ significantly from today's EELV, thus obviating the supposed advantage of commonality with DOD systems for our nation's launch vehicle industrial base. Finally, the transition from the shuttle to the Ares launch vehicle family is less disruptive for our workforce and makes more efficient use of existing facilities and ground support equipment than an EELV-derived system.

These are not trivial issues. These facts matter. And while I appreciate that many proponents of EELV systems were upset with our decision and some still are, I stand by it. We are following the data not opinion, emotion or a course of action based upon any personal benefit.

Turning now from the launch vehicle architecture to the overall Exploration architecture, NASA's Constellation systems are designed for the moon but must support the space station as well. Thus, we are sizing the Orion crew capsule and Ares rockets appropriately for both missions. Now, from numerous conversations with people who are genuinely friends and colleagues, I completely understand that some would prefer to replace the shuttle with a new system to support the space station and nothing more. They are either uninterested in venturing beyond Earth orbit or regard it as a problem for another generation. In my view, such a narrow focus would (again) leave NASA and our nation stuck in low Earth orbit. This is not the direction provided to NASA by President Bush and Congress in two successive authorization bills.

While some pundits have opined that we will receive new direction from a future president or Congress, we will continue to follow the law of the land as it exists today unless and until such new guidance is provided. I, for one, devoutly hope that we do not reverse course. Let me be very blunt: We have spent the last 35 years conducting the experiment of confining our ambitions for human spaceflight to low Earth orbit. It did not turn out in a manner befitting a great nation. Let's not continue it.

Finally, I would like to speak openly and honestly about the criticism that NASA did not study space access carefully enough in our 2005 Exploration Systems Architecture Study (ESAS). So I'll put it on the public record again: We conducted a thorough study by engineers who have considered this technical challenge for many years. The ESAS was the culmination, not the beginning, of these studies. Further, when and as we learned more after the ESAS we continued to incorporate new ideas, resulting in beneficial changes. For example, we eliminated over $5 billion in life cycle costs by adopting the RS-68 core engine and going directly to a common five-segment solid rocket booster and J-2X upper stage engine for both crew and heavy-lift launch vehicles.

The key cost and safety advantages of the shuttle-derived launch architecture remain as clear today as in 2005. If someone has better data or specific examples where the data I reviewed either in 2005 or since has been "skewed" or is incomplete, please come forward. I receive a lot of e-mails on this subject but none offering new or better data—only conjecture and opinion. To date, on the rare occasions when data is offered to support contrarian opinions, I have found it to ignore engineering reality, funding constraints or the law of the land. Finally, I would add that if we are stifling dissent then we are doing an extremely poor job of it, given the amount of ink provided to so many dissenters.

I am not putting my thumb on the scales. I believe our leadership team upholds the philosophy that we strive always to be receptive to constructive criticism in solving a problem. However, we are now well past the time when we can simply "stop work" to conduct more architecture studies. In my opinion, the propensity to conduct too many studies with too little action has in recent years been a profoundly detrimental characteristic of this nation's broader aerospace enterprise. It is long past time to *do* something. We are deeply engaged in the design, development and testing of the Orion and Ares I and we will be ramping up our work on the Ares V and Altair in the months ahead. We're making good progress. Let's keep it up.

We have—I have—explained quite carefully over the past several years why our new spaceflight architecture looks the way it does with our eyes wide open to the fact that this transition to a new system for human spaceflight transport will be the greatest challenge NASA has faced in decades. Now with all that said, if you have a better idea that doesn't conveniently ignore one or more of our many constraints, we'll listen. But bring data and be prepared to have the technical discussion. Simply saying that an idea is better does not make it so.

Now, please do not infer from my comments that I believe we have a perfect answer to the problems facing our nation's human spaceflight program. We don't. Our solution is simply the best we can construct with the funding provided to meet our long-standing commitment to complete the assembly of the space station while building new ships to embark on new ventures beyond low-Earth orbit.

It has only been 5 years since the report of the *Columbia* Accident Investigation Board and we have made tremendous strides as an organization in that time. It is my sincere hope that we have learned the lessons of past mistakes, taken them to heart and emerged stronger from that adversity. We did so after the Apollo 1 fire; we did it when *Challenger* was lost; and we can do it again now. NASA celebrates its 50th anniversary this year. We should celebrate our achievements but we must also remember those days when we failed to meet our own expectations. If we are to act with integrity, we must remember these failures by learning from them.

Dealing with failure is the essence of engineering. In his book, *To Engineer is Human: The Role of Failure in Successful Design*, Professor Henry Petroski notes, "No one wants to learn by mistakes, but we cannot learn enough from successes to go beyond the state of the art." Petroski expands greatly upon that theme in other works, including *Success Through Failure: The Paradox of Design*. If you have not read these works, I highly recommend them to you. You will come away from them with the understanding that, as unfortunate as it may be, most of the great advances in engineering have been the result of learning from failures. That willingness to learn is a sign of integrity.

I too am eager to learn of better ways by which NASA can accomplish its missions. However, I cannot learn from proposals that in the end come down to saying what NASA should or could do if we had more money. While I too personally wish that NASA had more money, it would be irresponsible for me not to be honest with our stakeholders at the White House and Congress about the careful balance we have reached with the resources provided. When the law and policy directives cannot be carried out within known funding limitations or defy our best engineering and management judgment, it is our duty to say so. We cannot simply do more with less. We should not over-promise and under-deliver. Saying "no" when that is the honest answer is also a sign of integrity.

In his book, *Good to Great*, Jim Collins evaluated how "good" companies become "great" companies. The primary task in becoming "great" is "to create a culture wherein people have a tremendous opportunity to be heard and, ultimately, for the truth to be heard." Collins further recommends that such "great" companies "face just as much adversity as [others], but respond to that adversity differently. They hit the realities of their situation head-on. As a result, they emerge from adversity even stronger."

The same holds true for high-performance government organizations like NASA. We need the best ideas to come forth, to learn from our experiences and for there to be civil dialogue and debate, not vituperation, before setting forth on the best course to follow.

The men and women of NASA are writing a new chapter in the history of space exploration. It's a complex story, a rich story, full of drama and despair, pride and pathos. It is a story we need to tell our children and grandchildren lest they forget why it is we explore what John F. Kennedy referred to as the New Frontier of space. I believe it is necessary for us to discuss openly and honestly the principles that led us as a nation to embrace space exploration five decades ago and the need to continue that journey. But first and foremost, we must tell our story with integrity.

Thank you.

Shown is a concept illustration of the Ares I crew launch vehicle (left) and Ares V cargo launch vehicle. Ares I will carry the Orion Crew Exploration Vehicle to space. Ares V will serve as NASA's primary large-scale hardware delivery vehicle. (Ares_Collage3)

At 9:32 a.m. EDT, the swing arms move away and a plume of flame signals the liftoff of the Apollo 11 Saturn V space vehicle and astronauts Neil A. Armstrong, Michael Collins and Edwin E. Aldrin, Jr. from Kennedy Space Center Launch Complex 39A. (GPN-2000-000629)

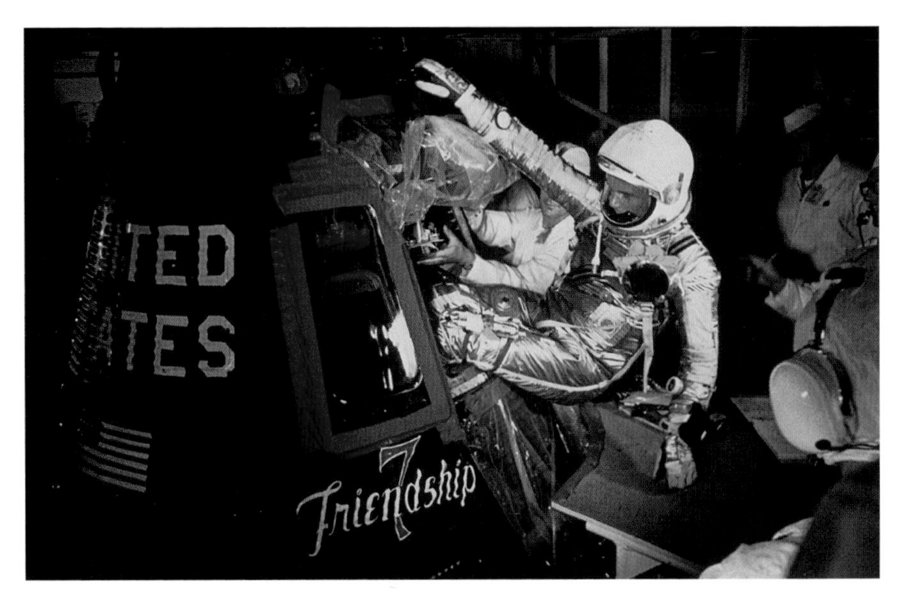

Astronaut John Glenn, Jr. as he enters into the spacecraft Friendship 7 prior to MA-6 launch operations at Launch Complex 14. Astronaut Glenn is entering his spacecraft to begin the first American manned Earth orbital mission. (GPN-2000-000652)

Atlas Agena target vehicle liftoff for Gemini 11 from Pad 14. Once the Agena was in orbit, Gemini 11 rendezvoused and docked with it. (GPN-2000-001019)

The Original Mercury Seven astronauts with a U.S. Air Force F-106B jet aircraft. From left to right: M. Scott Carpenter, Leroy Gordon Cooper, John H. Glenn, Jr., Virgil I. Gus Grissom, Jr., Walter M. Wally Schirra, Jr., Alan B. Shepard, Jr., Donald K. Deke Slayton. (GPN-2000-001286)

NASA's Constellation Program is getting to work on the new spacecraft that will return humans to the moon and be capable of delivering crews and cargo to the space station. The new spacecraft will be similar in shape to the Apollo spacecraft but significantly larger. The tried-and-true conical form is the safest and most reliable for re-entering Earth's atmosphere. This artist's rendering represents a concept of the Orion Crew Exploration Vehicle in orbit with the moon and Earth in the background. (jsc2008e097002)

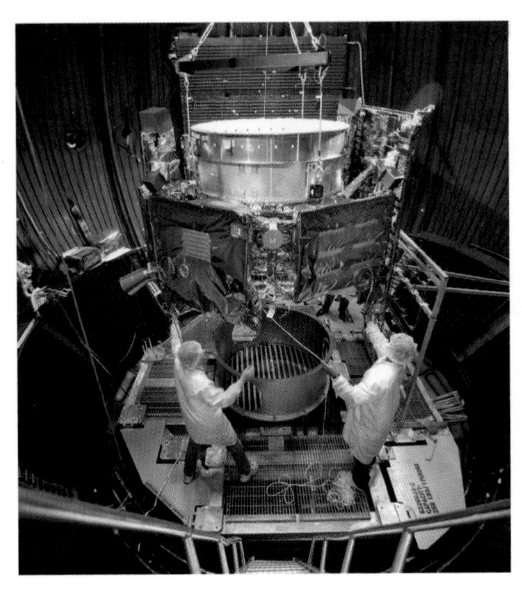

The LCROSS is gently maneuvered into the thermal vacuum chamber at the Northrop Grumman facility, Redondo Beach, California. The spacecraft was subjected to heating and cooling cycles to simulate the harsh conditions in outer space. Thermal vacuum is one of the final milestones before the LCROSS is certified for spaceflight. (LCROSS)

This image shows NASA's Phoenix Mars Lander's solar panel and the lander's robotic arm with a sample in the scoop. The image was taken by the lander's surface stereo imager looking west during Phoenix's Sol 16 (June 10, 2008) or the 16th Martian day after landing. The image was taken just before the sample was delivered to the optical microscope. This view is a part of the "mission success" panorama that will show the whole landing site in color. (PHOENIX_PANELARM)

Astronaut Edward H. White II, pilot for the Gemini-Titan 4 spaceflight, during America's first spacewalk. The EVA was performed during the Gemini 4 mission on June 3, 1965. White spent 23 minutes maneuvering while attached to the spacecraft by a 25-foot umbilical line and a 23-foot tether line, both wrapped in gold tape to form one cord. White carries a hand-held self maneuvering unit (HHSMU) to help move him in the weightless space environment. His helmet visor is gold plated to protect him from unfiltered sun rays. (S65-34635)

(1992) — Laser power stations, perhaps drawing energy from the local environment, might one day propel spacecraft. NASA studies of advanced planetary missions have ranged from small robotic probes to multiple-spacecraft human exploration missions. Credit: Pat Rawlings (SAIC), titled "Via Luminae." (S99-04188)

(1997) — A few kilometers from the Apollo 17 Taurus Littrow landing site, a lunar mining facility harvests oxygen from the resource-rich volcanic soil of the eastern Mare Serenitatis. Credit: Pat Rawlings (SAIC), titled "The Deal." (S99-04195)

Part 3.
Getting There from Here

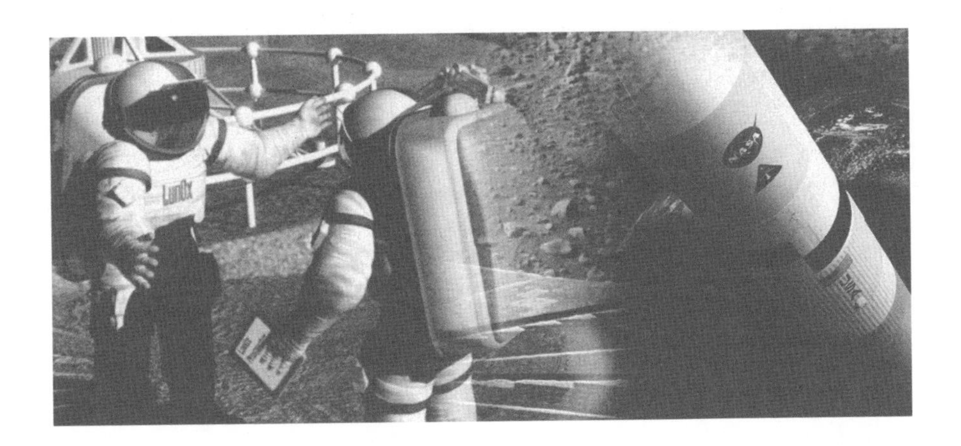

The Constellation Architecture

Michael D. Griffin
Administrator
National Aeronautics and Space Administration
Remarks to the Space Transportation Association
January 22, 2008

I've received multiple inquiries lately, in one forum or another, concerning various aspects of NASA's post-shuttle spaceflight architecture. None of the questions is new, and all of them were elucidated during our Exploration Systems Architecture Study (ESAS). The architecture is essentially as it was coming out of ESAS back in September 2005, and the architectural trades we made then when considering mission requirements, operations concepts, performance, risk, reliability and cost hold true today.

But more than 2 years have gone by, and the logic behind the choices we made has receded into the background. People come and go; new questioners lacking subject matter background appear; and the old questions must be answered again if there is to be general accord that NASA managers are allocating public funds in a responsible fashion. And so it seemed to me that the time was right to review again why we are developing the post-shuttle space architecture in the way that we are.

As many of you know, I used to teach space system engineering at The George Washington University and the University of Maryland and am more comfortable discussing engineering design than just about any other topic. But as NASA administrator, I must first frame the Constellation architecture and design in the context of policy and law that dictate NASA's missions.

Any system architecture must be evaluated first against the tasks that it is supposed to accomplish. Only afterward can we consider whether it accomplishes them efficiently or presents other advantages that distinguish it from competing choices. So to start, we need to review the requirements expressed in presidential policy and, subsequently, congressional direction that were conveyed to NASA in 2004 and 2005.

The principal documents pertinent to our architecture are President Bush's January 14, 2004 speech outlining the Vision for Space Exploration and the NASA Authorization Act of 2005. Both documents are a direct result of the policy debate that followed in the wake of the *Columbia* tragedy 5 years ago and the observation of the *Columbia* Accident Investigation Board (CAIB). "The U.S. civilian space effort has moved forward for more than 30 years without a guiding vision."

Several items of specific direction are captured in the president's speech:

- "Our first goal is to complete the International Space Station by 2010. We will finish what we have started; we will meet our obligations to our 15 international partners on this project."
- "Research onboard the station and here on Earth will help us better understand and overcome the obstacles that limit exploration. Through

these efforts we will develop the skills and techniques necessary to sustain further space exploration."

- "Our second goal is to develop and test a new spacecraft, the crew exploration vehicle … and to conduct the first manned mission no later than 2014. The crew exploration vehicle will be capable of ferrying astronauts and scientists to the space station after the shuttle is retired. But the main purpose of this spacecraft will be to carry astronauts beyond our orbit to other worlds."
- "Our third goal is to return to the moon by 2020. …"
- "With the experience and knowledge gained on the moon, we will then be ready to take the next steps of space exploration: human missions to Mars and to worlds beyond."

After extensive debate, Congress offered strong bipartisan approval of these goals, while adding considerable specificity. From the 2005 Authorization Act for NASA:

> The administrator shall establish a program to develop a sustained human presence on the moon, including a robust precursor program, to promote exploration, science, commerce and United States preeminence in space and as a stepping-stone to future exploration of Mars and other destinations.
>
> The administrator shall manage human spaceflight programs to strive to achieve the following milestones:

(A) Returning Americans to the moon no later than 2020.

(B) Launching the crew exploration vehicle as close to 2010 as possible.

(C) Increasing knowledge of the impacts of long duration stays in space on the human body using the most appropriate facilities available, including the space station.

(D) Enabling humans to land on and return from Mars and other destinations on a timetable that is technically and fiscally possible."

The bill establishes specific requirements for the space station, noting that it must "have an ability to support a crew size of at least six persons," codifying a long-promised design feature in law. It also details statutory requirements for shuttle transition, including maximizing the use of shuttle assets and infrastructure:

> The administrator shall, to the fullest extent possible consistent with a successful development program, use the personnel, capabilities, assets and infrastructure of the space shuttle program in developing the crew exploration vehicle, crew launch vehicle and a heavy-lift launch vehicle.

Collectively, these requirements outline the broad policy framework for the post-shuttle U.S. human spaceflight architecture:

1. We will manage the U.S. space program so as to complete the space station by 2010, utilizing the space shuttle for that purpose after which it will be retired.

2. After completion, the space station will be used to "better understand and overcome the obstacles that limit exploration."

3. The shuttle will be replaced as soon as possible, but not later than 2014, by a crew exploration vehicle designed to take humans to the moon and beyond, but which must also be capable of servicing the space station and its crew of six.

4. The architecture must support human lunar return not later than 2020 and, after that, development of a sustained human lunar presence, both for its intrinsic benefits and as a "stepping stone" to Mars and beyond.

5. Finally, the new architecture must take advantage of existing space shuttle program assets "to the fullest extent possible."

Not that anyone asked, but I consider this to be the best civil space policy to be enunciated by a president and the best authorization act to be approved by Congress since the 1960s. But no policy is perfect and none will please everyone. In particular, many in the exploration community as well as many of those who pursue space science were disappointed by the reaffirmation of our nation's commitment to the space station.

But a plain reading of policy and law requires us to understand that, throughout four presidential administrations and 20-plus congressional votes authorizing tens of billions of dollars for its development, the space station has remained an established feature of U.S. space policy. Its support and sustenance cannot be left to chance; the crew exploration vehicle must and will be capable of fulfilling this requirement; and the exploration architecture must and will take that into account. This is nothing more than common sense. The U.S. government will not abandon its commitment to the development and utilization of low Earth orbit.

There continues to be many questions about NASA's long-term commitment to the space station, so let me clarify. The Bush administration has made no decision on the end date for space station operations. We are, of course, concerned that station operating costs after 2016 will detract from our next major milestone: returning to the moon by 2020. But while the budget does not presently allocate funds for operating the space station beyond 2016, we are taking no action to preclude it. Decisions regarding U.S. participation in space station operations after 2016 can only be made by a future administration and a future Congress. I am sure these will be based on discussions with our international partners, progress toward our exploration goals, utility of this national laboratory and the affordability of projected space station operations.

Again, we plan to keep our commitments to our partners, utilizing the space station if it makes sense.

Now, returning to our space architecture, note the order of primacy in requirements. We are not primarily building a system to replace the shuttle for access to low Earth orbit and upgrading it later for lunar return. Instead, we are directed to build a system to "carry astronauts beyond our orbit to other worlds," but which can be put to the service of the space station if needed. In brief, we are designing for the moon and beyond.

That too is only common sense. Once before, an earlier generation of U.S. policymakers approved a spaceflight architecture intended to optimize access to low Earth orbit. It was expected—or maybe "hoped" is the better word—that with this capability in hand, the tools to resume deep space exploration would follow. It didn't happen. And with the funding which has been allocated to the U.S. civil space program since the late 1960s, it cannot happen. Even though from an engineering perspective it would be highly desirable to have transportation systems separately optimized for low Earth orbit and deep space, NASA's budget will not support it. We get one system; it must be capable of serving in multiple roles and it must be designed for the more difficult of those roles from the outset.

There are other common-sense requirements that have not been written down.

The most obvious of these, to me, is that the new system will and should be in use for many decades. Aerospace systems are expensive and difficult to develop; when such developments are judged successful, they tend to remain in use far longer than one might at first imagine. Those who doubt this should look around. The DC-3 and the B-52, to name only two landmark aircraft, remain in service today. The Boeing 747 has been around for 30 years and who doubts that it will be going strong for another 30? In space, derivatives of Atlas, Delta and *Soyuz* are flying a half-century and more after their initial development. Ariane and its derivatives have been around for three decades with no end in sight. Even the space shuttle will have been in service for 30 years by the time it retires. Apart from Saturn and Apollo, I am hard put to think of a successful aerospace system which was retired with less than several decades of use and often more.

The implications of this are profound. We are designing today the systems that our grandchildren will use as building blocks, not just for lunar return but for missions to Mars, to the near-Earth asteroids, to service great observatories at sun-Earth L1 and for other purposes we have not yet even considered. We need a system with inherent capability for growth.

Elsewhere, I have written that a careful analysis of what we can do at NASA on constant-dollar budgets leads me to believe that we can realistically be on Mars by the mid-2030s. It is not credible to believe that we will return to the moon and then start with a "clean sheet of paper" to design a system for Mars. That's just not fiscally, technically or politically realistic. We'll be on Mars in 30 years; and when we go, we'll be using hardware that we're building today.

So we need to keep Mars in mind as we work (even now). And that means we need to look at both ends of the requirements spectrum. Our new system needs to be designed for the moon but allow the U.S. government access to low Earth orbit. Yet, in designing for the moon, we need also to provide the maximum possible "leave behind" for Mars. If we don't, then a generation from now there will be a group in this room listening to the administrator of that time ask about those of us here today: "What were they thinking?"

Now, in mentioning "Mars" I must state for the record that I do realize that the $550 billion Consolidated Appropriations Act signed into law last month stipulated that no funds appropriated in 2008 "shall be used for any research, development or demonstration activities related exclusively to the human exploration of Mars." While I personally consider this to be shortsighted, and while NASA was in any case spending only a few million dollars on long-term research and study efforts, we will of course follow this legislative direction. And while this provision does not affect work on Ares V, it does call into question the fundamental rationale for our use of space station in long-duration human spaceflight research. I hope that this funding restriction can be abandoned in future years.

Further application of common sense also requires us to acknowledge that now is the time; this is the juncture; and we are the people to make provisions for the contributions of the commercial space sector to our nation's overall space enterprise. The development and exploitation of space has, so far, been accomplished in a fashion that can be described as "all government, all the time." That's not the way the American frontier was developed; it's not the way this nation developed aviation; it's not the way the rest of our economy works; and it ought not to be good enough for space, either. So, proactively and as a matter of deliberate policy, we need to make provisions for the first step on the stairway to space to be occupied by commercial entrepreneurs whether they reside in big companies or small ones.

The policy decision that the crew exploration vehicle will be designed for the moon, while not precluding its ability to provide access to low Earth orbit, strongly reinforces this common sense objective. If designed for the moon, the use of the crew exploration vehicle in low Earth orbit will inevitably be more expensive than a system designed for the much easier requirement of low Earth orbit access and no more. This lesser requirement is one that, in my judgment, can be met today by a bold commercial developer operating without the close oversight of the U.S. government with the goal of offering transportation for cargo and crew to low Earth orbit on a fee-for-service basis.

This is a policy goal—enabling the development of commercial space transportation to low Earth orbit—that can be met if we in government are willing to create a protected niche for it. To provide that niche, we must set the requirements for the next-generation government spaceflight system at the lunar-transportation level well above the low Earth orbit threshold.

Now again, common sense dictates that we cannot hold the space station hostage to fortune; we cannot gamble the fate of a multi-tens-of-billions-of-dollars facility on the success of a commercial operation; so the crew exploration vehicle must be able to operate efficiently in low Earth orbit if necessary. But we can create a clear financial incentive for commercial success based on the financial disincentive of using government transportation to low Earth orbit at what will be an inherently higher price.

To this end (as I have noted many times) we must be willing to defer the use of government systems in favor of commercial services as and when they reach maturity. When commercial capability comes on line, we will reduce the level of our own low Earth orbit operations with Ares and Orion to that which is minimally necessary to preserve capability and to qualify the systems for lunar flight.

So how is all of this—law, policy and common sense—realized in the architecture that came out of ESAS?

As I have outlined above, policy and legislation are in some ways quite specific about the requirements for post-shuttle U.S. spaceflight systems. They are less so where it concerns our lunar goals, beyond the clearly stated requirement to develop the capability to support a sustained human lunar presence both for its intrinsic value and as a step toward Mars. This leaves considerably more discretion to NASA as the executive agency to set requirements and with that considerably more responsibility to get it right. Again, I think common sense comes to our rescue.

There is general agreement that our next steps to the moon, toward a goal of sustained lunar presence, must offer something more than Apollo-class capability; for example, sorties by two people for 3 days to the equatorial region.

To return after 50 years with nothing more than the capability we once threw away seems to me to fail whatever test of common sense might be applied to ourselves and our successors.

Accordingly, then, in developing requirements for ESAS we specified that the lunar architecture should be capable of the following:

- Initial lunar sortie missions should be capable of sustaining a crew of four on the lunar surface for a week.
- The architecture will allow missions to any location on the moon at any time and will permit return to Earth at any time.
- The architecture will be designed to support the early development of an "outpost" capability at a location yet to be specified, with crew rotations planned for 6-month intervals.

One could fill pages debating and justifying these requirements; mercifully, I will not do that. Perhaps another time. In any case, I think it is clear that these goals offer capability significantly beyond Apollo. Yet they can be achieved with the building blocks—ground facilities as well as space transportation elements—that we have or can reasonably envision given the budgetary resources we might expect.

It is worth noting that the decision to focus on early development of an outpost—while retaining the capability to conduct a dedicated sortie mission to any point on the lunar surface that might prove to be of interest for scientific or other reasons—supports additional key goals. The most obvious of these is that it provides a more direct "stepping stone" to Mars, where even on the very first mission we will need to live for an extended period on another planetary surface. But further, even a basic human-tended outpost requires a variety of infrastructure that is neither necessary nor possible to include in a sortie mission. Such infrastructure development presents obvious possibilities for commercial and international partner involvement, both of which constitute important policy objectives.

But if the capability we are striving for is greater than that of Apollo, so too is the difficulty. To achieve the basic four-person lunar sortie capability anytime, anywhere requires a trans-lunar injection mass of 70–75 metric tons, including appropriate reserve. Saturn V trans-lunar injection mass capability on Apollo 17 was 47 metric tons without the launch adaptor used to protect the lunar module. Thus, more than Saturn V capability is required if we are to go beyond Apollo. I think we should not be surprised to find that the Apollo engineers got just about as much out of a single launch of the Saturn V as it was possible to do.

If we need more trans-lunar injection mass capability than can be provided by a single launch of a Saturn-class vehicle, we can reduce our objectives, build a bigger rocket or attain the desired capability by launching more than one rocket. Setting a lesser objective seems inconsistent with our goal of developing the capability for a sustained lunar presence and, as noted earlier, merely replicating Apollo-era capability is politically untenable.

Building a larger rocket is certainly an attractive option, at least to me; but to reach the capability needed for a single launch brings with it the need for significant modifications to fabrication and launch infrastructure. The Michoud Assembly Facility and the Vertical Assembly Building were designed for the Saturn V and have some growth margin above that. But they will not accommodate a vehicle that can support our goals for lunar return with a single launch; and the projected NASA budget does not allow the development of extensive new ground infrastructure. Further, and crucially, a single-launch architecture fails to address the requirement for space station logistics support.

Thus, after detailed consideration of the single-launch option, we settled on a dual-launch Earth-orbit rendezvous scheme as the means by which a trans-lunar injection mass payload of the necessary size would be assembled.

However, the decision to employ Earth-orbit rendezvous in the lunar transportation architecture implies nothing about how the payload should be split. Indeed, the most obvious split involves launching two identical vehicles with approximately equal payloads, mating them in orbit and proceeding to the moon. When Earth-orbit rendezvous as considered for Apollo, it was this method that was to be employed; and it offers several advantages. Non-recurring costs are lower because only one launch vehicle development is required; recurring costs are amortized over a larger number of flights of a single vehicle; and the knowledge of system reliability is enhanced by the more rapid accumulation of flight experience.

However, this architectural approach carries significant liabilities when we consider the broader requirements of the policy framework discussed earlier. As with the single-launch architecture, dual-launch Earth-orbit rendezvous of identical vehicles is vastly overdesigned for space station logistics. It is one thing to design a lunar transportation system and, if necessary, use it to service the space station while accepting some reduction in cost-effectiveness relative to a system optimized for low Earth orbit access. As noted earlier, such a plan backstops the requirement to sustain the space station without offering government competition in what we hope will prove to be a commercial market niche. But it is quite another thing to render government logistics support to the space station so expensive that the station is immediately judged not to be worth the cost of its support. Dual-launch Earth-orbit rendezvous with vehicles of similar payload class does not meet the requirement to support the space station in any sort of cost-effective manner.

On the other end of the scale, we must judge any proposed architecture against the requirements for Mars. We aren't going there now, but one day we will and it will be within the expected operating lifetime of the system we are designing today. We know already that, when we go, we are going to need a Mars ship with a low Earth orbit mass equivalent of about 1 million pounds, give or take a bit. I'm trying for one-significant-digit accuracy here but think "space station" in terms of mass.

I hope we're smart enough that we never again try to place such a large system in orbit by doing it in 20-ton chunks. I think we all understand that fewer launches of larger payloads requiring less on-orbit integration are to be preferred. Thus, a vehicle in the Saturn V class—some 300,000 pounds in low Earth orbit—allows us to envision a Mars mission assembly sequence requiring some four to six launches depending on the packaging efficiency we can attain.

This is something we did once and can do again over the course of a few months, rather than many years, with the two heavy-lift pads available at Kennedy Space Center Complex 39. But if we split the Earth-orbit rendezvous lunar architecture into two equal but smaller vehicles, we will need 10 or more launches to obtain the same Mars-bound payload in low Earth orbit, and that is without assuming any loss of packaging efficiency for the launch of smaller payloads. When we consider that maybe half the Mars mission mass in low Earth orbit is liquid hydrogen, and if we understand that the control of hydrogen boiloff in space is one of the key limiting technologies for deep space exploration, the need to conduct fewer rather than more launches to low Earth orbit for early Mars missions becomes glaringly apparent.

So if we want a lunar transportation architecture that looks back to the space station low Earth orbit logistics requirement and forward to the first Mars missions, it becomes apparent that the best approach is a dual-launch Earth-orbit rendezvous mission with the total payload split unequally. The smaller launch vehicle puts a crew in low Earth orbit every time it flies, whether they are going to the space station or the moon. The larger launch vehicle puts the lunar (or later Mars) cargo in orbit. After rendezvous and docking, they are off to their final destination.

Once the rationale for this particular dual-launch Earth-orbit rendezvous scenario is understood, the next question is logically, "why don't we use the existing Evolved Expendable Launch Vehicle (EELV) fleet for the smaller launch?" I'm sure you will understand when I tell you that I get this question all the time. And frankly, it's a logical question. I started with that premise myself some years back. To cut to the chase, it will work as long as you are willing to define "Orion" as that vehicle which can fit on top of an EELV. Unfortunately, we can't do that.

The adoption of the shuttle-derived approach of Ares I, with a new lox/hydrogen upper stage on a reusable solid rocket booster (RSRB) first stage, has been one of our more controversial decisions. The Ares V heavy-lift design, with its external-tank-derived core stage augmented by two RSRBs and a new Earth departure stage has been less controversial but still not without its detractors. So let me go into a bit of detail concerning our rationale for the shuttle-derived approach.

The principal factors we considered were the desired lift capacity, the comparative reliability and the development and lifecycle costs of competing approaches. Performance, risk and cost—I'm sure you are shocked.

The Ares I lift requirement is 20.3 metric tons for the space station mission and 23.3 metric tons for the lunar mission. Evolved Expendable Launch Vehicle lift capacity for both the Delta IV and Atlas V are insufficient, so a new RL-10 powered upper stage would be required, similar to the J-2X based upper stage for Ares I. We considered using additional strap-on solid rocket boosters to increase EELV performance, but such clustering lowers overall reliability.

It is also important to consider the growth path to heavy lift capability, which results from the choice of a particular launch vehicle family. Again, we are designing an architecture, not a point solution for access to low Earth orbit. To grow significantly beyond today's EELV family for lunar missions requires essentially a "clean sheet of paper" design, whereas the Ares V design makes extensive use of existing elements or straightforward modifications of existing elements, which are also common to Ares I.

Next up for consideration are mission reliability and crew risk. EELVs were not originally designed to carry astronauts, and various human-rating improvements are required to do so. Significant upgrades to the Atlas V core stage are necessary and abort from the Delta IV exceeds allowable g-loads. In the end, the probabilistic risk assessment derived during ESAS indicated that the shuttle-derived Ares I was almost twice as safe as that of a human-rated EELV.

Finally, we considered both development and full lifecycle costs. I cannot go into the details of this analysis in a speech; and in any case, much of it involves proprietary data. We have shared the complete analysis with the Department of Defense (DOD), various White House staff offices, the Congressional Budget Office (CBO), the Government Accountability Office (GAO), and our congressional oversight committees. Our analysis showed that for the combined crew and heavy-lift launch vehicles, the development cost of an EELV-derived architecture is almost 25 percent higher than that of the shuttle-derived approach. The recurring cost of the heavy-lift Ares V is substantially less than competing approaches, and the recurring cost of an EELV upgraded to meet CEV requirements is, at best, comparable to that for Ares I. All independent cost analyses have been in agreement with these conclusions.

So, while we might wish that "off the shelf" EELVs could be easily and cheaply modified to meet NASA's human spaceflight requirements, the data say otherwise. Careful analysis showed EELV-derived solutions meeting our performance requirements to be less safe, less reliable and more costly than the shuttle-derived Ares I and Ares V.

Now is a good time to recall that all of the trades discussed above assumed the use of a production version of the space shuttle main engine (SSME). But, returning to a point I made earlier, we continued our system analysis following the architecture definition of ESAS, looking for refinements to enhance performance and reduce risk and cost. We decided for Ares I to make an early transition to the five-segment reusable solid rocket booster and to eliminate the SSME in favor of the J-2X on the upper stage. Similarly, elimination of the SSME in favor of an upgraded version of the Air Force-developed RS-68 engine for the Ares V core stage with the Earth departure stage powered by the J-2X offered numerous benefits. These changes yielded several billions of dollars in lifecycle cost savings over our earlier estimates and foster the use of a common RS-68 core engine line for DOD, civil and commercial users.

Praise is tough to come by in Washington, so I was particularly pleased with the comment about our decision on the five-segment RSRB and J-2X engine in the recent GAO review: "NASA has taken steps toward making sound investment decisions for Ares I." Just for balance, of course, the GAO also provided some other comments. So, for the record, let me acknowledge on behalf of the entire Constellation team that, yes, we do realize that there remain "challenging knowledge gaps," as the GAO so quaintly phrased it, between system concepts today and hardware on the pad tomorrow. Really. We do.

It's time now for a little perspective. We are developing a new system to bring new capabilities to the U.S. space program, capabilities lost to us since the early 1970s. It isn't going to be easy. Let me pause for a moment and repeat that: It isn't going to be easy.

So no, we don't yet have all the answers to the engineering questions we will face. And in some cases we don't even know what those questions will be. That is the nature of engineering development. But we are going to continue to follow the data in our decision-making, continue to test our theories and continue to make changes if necessary.

We have been, I think, extraordinarily open about all of this. Following the practice I enunciated in my first all-hands on my first day as administrator, in connection with the then-pressing concerns about shuttle return-to-flight, we are resolved to listen carefully and respectfully to any technical concern or suggestion which is respectfully expressed and to evaluate on their merits any new ideas brought to us. We are doing that every day. We will continue to do it.

So, in conclusion, this is the architecture which I think best meets all of the requirements of law, policy, budget and common sense that constrain us in the post-shuttle era. It certainly does not satisfy everyone; not that I believe that goal to be achievable. To that point, one of the more common criticisms I receive is that it "looks too much like Apollo." I'm still struggling to figure out why it is bad (if indeed that is so).

My considered assessment of the Constellation architecture is that while we will encounter a number of engineering design problems as we move forward, we are not facing any showstoppers. Constellation is primarily a systems engineering and integration effort based on the use of as many flight-proven concepts and hardware as possible, including the capsule design of Orion, the shuttle RSRBs and external tank, the Apollo-era J-2X upper stage engine and the RS-68 core engine. We're capitalizing on the nation's prior investments in space technology wherever possible. I am really quite proud of the progress this multi-disciplinary, geographically dispersed, NASA/industry engineering team has made thus far.

But even so, the development of new systems remains hard work. It is not for the faint of heart, or those who are easily distracted. We can do it if—but only if—we retain our sense of purpose.

In this regard, I'm reminded of two sobering quotes from the CAIB report. First, "the previous attempts to develop a replacement vehicle for the aging shuttle represent a failure of national leadership." Also, the CAIB noted that such leadership can only be successful "if it is sustained over the decade; if by the time a decision to develop a new vehicle is made there is a clearer idea of how the new transportation system fits into the nation's overall plans for space; and if the U.S. government is willing at the time a development decision is made to commit the substantial resources required to implement it."

That sort of commitment is what the mantle of leadership in space exploration means, and the engineers working to build Constellation know it every day. Thus, I can only hope to inspire them, and you, with the immortal words of that great engineer, Montgomery Scott, of the *USS Enterprise*: "I'm givin' 'er all she's got, Captain."

Thank you.

Then and Now: Fifty Years in Space

Michael D. Griffin
Administrator
National Aeronautics and Space Administration
AIAA Space 2007 Conference
Long Beach, California
September 19, 2007

The American Institute of Aeronautics and Astronautics (AIAA) has always been and will always be my primary professional society, so I believe I should start by paying respect to the wishes of our executive director and my longtime colleague, Bob Dickman. Bob has asked a number of us to recount the story of "when we knew;" that is, when it was that we knew that the aerospace profession would become our future. I think that too many of you have heard my own story already and I am a bit reluctant to bore you with it again, but maybe there is some value in getting it firmly into the record. Bob certainly thinks so; and so I'll recount it again.

Many of you have known me for a long time, and so you will not be surprised to hear that I was an unusual child, quite different from my parents, siblings and others of our acquaintance. There are pictures of me as a pre-kindergarten child trying to use my father's tools; and one of my very earliest recollections is of begging, each year, for my parents to get me an Erector Set for Christmas. This finally arrived, if I recall correctly, when I was in the second grade. I thought I had gone to heaven. If ever anyone was born to be an engineer, I was that person.

In 1954 or 1955, when I was 5 or 6 years old, my mother gave me as a birthday or Christmas present the first book that I can specifically recall receiving. It was entitled *A Child's Book of Stars*. Today, we know that most of what is in that book is wrong; but across more than five decades, with the vivid clarity of some childhood memories that each of us has, I can recall how utterly fascinated I was with it. I read and re-read it until I had virtually memorized it. I marveled that Halley's Comet would return in 1986 when I would be a whole 37 years old—I couldn't imagine such a thing. I began seeking out other books and magazine articles having anything to do with space and spaceflight. Many people have noted how reading science fiction as a child got them interested in space and other technical careers. For me it was the reverse; I began reading science fiction because I was interested in space!

As a small child, I had no clear understanding of the difference between mathematics, science and engineering; and I didn't care. I'm not sure that I really care all that much today. I did know that, whatever I did, it was going to have something to do with space. As time went by, I fell more broadly in love with the beauty of mathematics and its ability to describe the system of the world, with science and its quest to understand that world and most of all with engineering and its fusion of mathematics, science and human artistry to create a world which had never been. But I remained in love with space and later with flight in all its forms; and so that is where I chose to try to create that world.

That one book and that one decision shaped my life, sometimes in ways not so obvious. For example, I turned 21 in 1970, and I was fortunate—or maybe not—to be able to attain acceptable grades in college without expending much effort on schoolwork. So, you might not be surprised when I tell you that, shall we say, I experienced the 1960s to their fullest. It was a turbulent time and some of my cohort did not make it through; that era produced a lot of shattered lives. But for me, there was always a final governor on my behavior. No matter what, I would not indulge in anything that would prevent me from graduating and taking my place in the aerospace profession.

As a footnote to this story, I still have the book, thanks to the wisdom of mothers in preserving things that their careless children would discard. In as big a surprise as I have ever had, Eileen Collins and her crew presented it to me after having flown it, without my knowledge, on STS-114. Today it is mounted on my wall, together with photographs and a certificate of authenticity from the crew attesting to the event of its flight aboard *Discovery*. I suppose that is as close as I will get to spaceflight.

This reverie reminds me that our society loves to celebrate milestones—birthdays, anniversaries or holidays that celebrate our heroes and heritage, that honor the great struggles our ancestors undertook and the sacrifices that they made in the service of our country or that recall landmark historical and cultural achievements. This year, we in the space business are celebrating 50 years of spaceflight dating from the launch of *Sputnik* on October 4, 1957. In the process, we honor its legacy and look back thoughtfully on its influence on the course of human events. I go to Russia next month to join them in celebration of that achievement.

Now, there are appropriate times and occasions for retrospection and this is one of them. But to me it is always more important to look forward; and I like to use such historical milestones to measure our progress and as a guide to setting challenging goals for the future. It is my hope that our greatest days lie always ahead of us. But with that said, another of my beliefs is that it is vitally important to learn from history, and the lessons available to Americans from our side of the *Sputnik* experience are no exception.

As anyone my age or thereabouts will recall personally, the Soviet Union's success with *Sputnik* was an almost unimaginable embarrassment for the United States. The reality of this event diverged impossibly from our image of ourselves and our place in the world. Notwithstanding the efforts and sacrifices of other nations, it was U.S. technical, industrial and logistical superiority which had been decisive in the then-recent global war. Other nations, victors and vanquished, were prostrate for a generation after that war. Japan and Germany were flattened. China and most of Europe were little better off. Fifty-five million people around the world were lost; the Soviet Union alone lost 20 million. The United States lost a few hundred thousand and our industrial infrastructure was untouched. Our nation stood like a colossus above the postwar world.

We had our own satellite development effort underway in support of the International Geophysical Year; putting an instrumented payload, the Vanguard satellite, into orbit was a publicly announced goal. The idea that we could be beaten to it by any other nation, not to mention by our declared adversary, was for almost everyone a paradigm-shifting event.

First and most directly, Americans felt vulnerable to Soviet missiles that, if they could place a payload in orbit, could also strike anywhere in the United States. In the generations since the founding of our nation, no other adversary had ever produced such a threat, and we had never imagined that anyone ever could. Nikita Krushchev's November, 1956 admonition—"We will bury you"—reverberated in our collective consciousness.

But there was more. In being beaten so publicly by what we then regarded as a peasant nation, a nation whose totalitarian government embraced a set of values abhorrent to nearly all Americans, we felt that we were falling behind in our much-vaunted technical know-how and industrial capability. The small metal orb beeping overhead, visible in the clear fall sky to anyone who looked—and nearly everyone did—reminded us of this. We felt that we were in second place in a new arena, that we lagged in exploring what President Kennedy later named so perfectly as "this new Ocean." And we felt that it mattered.

Given the story I related earlier, you will not be surprised to know that even at the age of 8 I followed these events avidly, scanning papers from *Weekly Reader* to *The Baltimore Sun* for news of space. I remember watching *Sputnik* from my home in Aberdeen, Maryland. I was hardly the only one doing so; as a nation, we looked up and contemplated the meaning of the Soviet accomplishment. The newspapers were full of both soul-searching analysis and opportunistic second-guessing. We questioned our military plans, our civilian research programs and our educational systems; we made changes in all those areas and more. America's readiness—or more properly our lack of readiness—to explore and exploit the space frontier decided a presidential election.

So began the Space Race. *Sputnik* changed everything. When the first Vanguard launch failed in December 1957, it spurred President Eisenhower to assign Army General John B. Medaris and his team, Wernher von Braun at Redstone Arsenal and William H. Pickering at the Jet Propulsion Laboratory (JPL) to build the Jupiter-C rocket and Explorer-1 satellite, matching the Soviet feat (4 months after the fact). It spurred the creation of NASA in October 1958 as well as the then-classified National Reconnaissance Office and its early CORONA satellites. It spurred Air Force General Bernard Schriever's team to develop the Atlas, Titan and Minuteman Intercontinental Ballistic Missiles (ICBMs). It spurred Admiral Red Raborn and his team to develop the Polaris missile and the Fleet Ballistic Missile Submarine to carry it. It spurred more national spending on science, math and engineering education to boost our nation's technical literacy, ultimately allowing us to win the Space Race and, eventually, the Cold War.

Millions of Americans, myself included, benefited from this investment of our nation's time, energy and resources; and this investment helped me personally and many others, even years later, to cut their teeth in the space business. We committed ourselves to the ideal that, as a nation, America must lead in space exploration. In President Kennedy's immortal words, prose but almost poetry, "For while we cannot guarantee that we shall one day be first, we can guarantee that any failure to make this effort will make us last." And, "We go into space because whatever mankind must undertake, free men must fully share."

Spaceflight is a strategic capability for our nation. We are fully 50 years into it now, measured from the launch of *Sputnik*. But equally, we are only 50 years into it. If you joined the space program after college graduation in

1958 or 1959, like my friends and mentors Glynn Lunney and Arnie Alrich, it all fits within a single long, full, tumultuous career. So we need to maintain a sense of perspective. Fifty years into the development of aviation, we didn't have even the first of the passenger-carrying jets that brought most of us to this conference. And 50 years into the development of open ocean seafaring, Carnival Cruise Lines was not one of the foreseeable outcomes. We have only just begun to sail the new ocean of space. We have a very long way yet to go. So let's spend a bit of time today to look back at where we have been, where we are and where we are going.

Fifty years ago, we in the United States were not yet aware that we were about to be surprised by *Sputnik*. We did not know that, little more than 1 year later, a new agency would be formed from elements of Navy, Army and National Advisory Committee for Aeronautics (NACA) laboratories. We did not know that seven test pilots would, in the spring of 1959, be named as the nation's first astronauts and would begin preparing for brief, single-seat trips into space. First and foremost, the goal was to beat the Russians to that prize. We all know how that turned out.

But times have changed. People change. Fifty years ago I looked up in the sky to watch *Sputnik* pass. I remember still the annoyance I felt, even at the age of 8, that we were not up there also. Today, two Russians and American astronaut Clayton Anderson are living and working in orbit aboard the International Space Station, the greatest engineering project in the brief history of the Space Age. Next month, NASA's Expedition 16 Commander Peggy Whitson, Russian Flight Engineer Yuri Malenchenko and spaceflight participant Sheikh Muszaphar Shukor, a Malaysian physician, will lift off from Baikonur to join them on the space station. Later in October, Space Shuttle *Discovery* and the STS-120 crew will launch from Kennedy Space Center carrying the Node 2 *Harmony* Module, built in Italy, which will be used to connect the European and Japanese laboratory modules. With the space station, NASA and our international partners have maintained a permanent human foothold in space since October 2000. We are learning to live and work in space 24/7, 365 days a year. We are engaged in a partnership with our former rival and even pay for some of their crew and cargo service to the space station. Naturally, I would rather be providing funding to U.S. commercial crew and cargo transportation providers, especially after we retire the space shuttle in 2010, but that is the position in which we find ourselves.

The space station is an engineering test bed and scientific laboratory to study and mitigate the hazards of long-duration spaceflight, just as sailors hundreds of years ago learned how to stave off the debilitating effects of scurvy and other hazards during long sea voyages. In the process, we will also develop new technologies that will improve life here on Earth. For example, the exacting techniques used for preserving food for our space missions have found their way into the Food and Drug Administration's safety standards. On our last shuttle mission, we flew a muscle atrophy experiment for the pharmaceutical company Amgen. Recently, a convention of the American Medical Association endorsed NASA's efforts in human spaceflight and the technologies and techniques we have developed for doctors. And last week, I signed a Memorandum of Understanding between NASA and the National Institutes of Health to conduct joint medical research onboard the space station.

Now, while I yield to no one in my belief that the advancement of the arts and sciences of human spaceflight is crucial to this nation, it is not the only measure of our progress over the last 50 years. Even our earliest robotic satellites produced truly lasting scientific results. Just over 50 years ago, Explorer 1 allowed James van Allen to infer the existence of the radiation belts circling Earth that now bear his name. And Vanguard 1, launched a few months later, showed that Earth was not actually round but in fact was a bit pear-shaped. NASA's work in space physics and Earth science continues today, albeit on a much grander scale, and we have added many new disciplines to our scientific repertoire in the meantime.

Forty-plus years ago as a teenager, I waited anxiously to see the first close-up photographs of the moon by Ranger 7 and a bit later went through the same nail-biting exercise as Mariner 4 executed the first Mars flyby. And I will simply never forget, as a young JPL engineer in the late 1970s, taking meals in the cafeteria while watching fresh new views of the moons of Jupiter from Voyager 1 and later Voyager 2 come up on the television screens. Today's planetary scientists can look forward to such new views of Mars every day, right from the surface, as Spirit and Opportunity continue to break new trails. And our presence extends out to Saturn where Cassini similarly rewards us with new views of Saturn and its moons every day. By 2011 we'll add Mercury—a very tough place to reach—to the list with the Mercury Surface, Space Environment, Geochemistry and Ranging (MESSENGER) spacecraft; and in 2015 we'll finally reach Pluto with New Horizons.

As a young flight controller at the Goddard Space Flight Center in the early 1970s, I had my virtual hands on the controls for the early radio astronomy explorers. Today, engineers and astrophysicists like NASA's John Mather are designing, building and controlling spacecraft that deal with issues that having metaphysical and religious overtones—the birth, life and death of stars, galaxies and our very universe.

Edwin Hubble discovered in the 1920s that the universe was expanding; in the late 1990s, astronomers using the Hubble Space Telescope discovered that the expansion of the universe was actually speeding up. Astrophysicists today explain such phenomena in terms of "dark matter" and "dark energy." This just means they can't see it and don't understand it. Recent observations by Hubble, combined with the European Very Large Telescope in Chile, Japan's Subaru telescope in Hawaii, the VLA radio telescope in New Mexico and the joint NASA-European Space Agency XMM-Newton satellite have revealed a loose network of filaments where normal matter in the form of galaxies accumulates along the densest concentrations of dark matter. That is the only way, so far, in which we know it's there. This unseen dark matter comprises approximately 20 percent of the mass-energy density of the observed universe while dark energy makes up maybe 75 percent of the universe. Everything we see, everything we know anything at all about, makes up the remaining 5 percent.

These are the deepest mysteries that there are, and it is heady stuff for someone like me, a simple aerospace engineer from a small town.

But the National Academy recognizes the significance of these issues and has set the Joint Dark Energy Mission (JDEM) as its highest priority for NASA's Beyond Einstein program. We support that priority and I look forward to working with Sam Bodman and Ray Orbach at the Department of Energy to turn this mission into reality. As with most space science missions NASA conducts, it is my hope that we will also have a great partnership on JDEM with the Centre National d'Etudes Spatiales (CNES), the French Space Agency. I discussed this with Yannick D'Escatha a few days ago during his recent visit to Washington, and I look forward to working together with him on other projects and programs as well.

International cooperation is the *sine qua non* for so many of NASA's science missions—astronomy and astrophysics, planetary science, heliophysics and Earth sciences—just as it is in our human spaceflight endeavors. It is through such work that we find our common humanity, discovering that what connects us is far more important than what divides us.

I believe that the art, science and business of engineering for space exploration is the hardest thing we do as a people. It is also the grandest expression of human imagination of which I can conceive. Members of the AIAA know this to be true. Just ask SpaceX, whom I will be visiting later for a progress update. Or ask the Northrop Grumman folks in Redondo Beach, who are building the James Webb Space Telescope; or those up the road at JPL in Pasadena where they are building the Mars Science Lab; or those at the Kennedy Space Center preparing to launch the Dawn mission to the asteroid belt next week. My hope is that when we contemplate the majesty of our universe resulting from complex instruments operating at a Lagrange point or landing on the surface of Mars, we appreciate the effort that went into making such technological miracles happen. They truly are miracles by the standards we used only 50 years ago.

But it is not only the miracles that matter. Less than 1 year from now, NASA will launch the final servicing mission to Hubble on Space Shuttle *Atlantis*. My hope is that members of the astronomy and astrophysics community will not only thank the astronauts who risk their lives in making this mission happen, but also that we all pause to recall the Hurricane Katrina ride-out crew who risked their lives to save the Michoud Assembly Facility near New Orleans where the space shuttle external tanks are manufactured. They were just as instrumental in making the Hubble servicing mission happen.

In a few years, this Michoud workforce will turn its attention to building the Ares 1 and Ares 5 rockets. While ESA's Ariane 5, which will launch the Webb Space Telescope, can loft approximately 21,000 kilograms to low-Earth orbit, the Ares V will be able to launch more than six times that amount. The Ares launch vehicles are being built primarily for human spaceflight, but my hope is that the engineers designing robotic spacecraft to be launched over the coming decades will take advantage of this new capability. Our nation has not seen such a capability since the Saturn 5 last flew in 1973; so it behooves us—not simply NASA but all of us in the space business—to begin now to think of innovative ways to use these launch systems for greater scientific and other benefits over the next 50 years.

We will soon be ready to discuss how we in NASA are organizing the workforce at our 10 field centers to lead and support development for our next major exploration programs, including those beyond Orion and Ares 1: the Ares V, the lunar surface access module and other lunar surface systems. This is the exploration work to be done over the next 10–15 years, and I hope to entice

Figure 1. GDP Inflated NASA Budget (1959–2008)

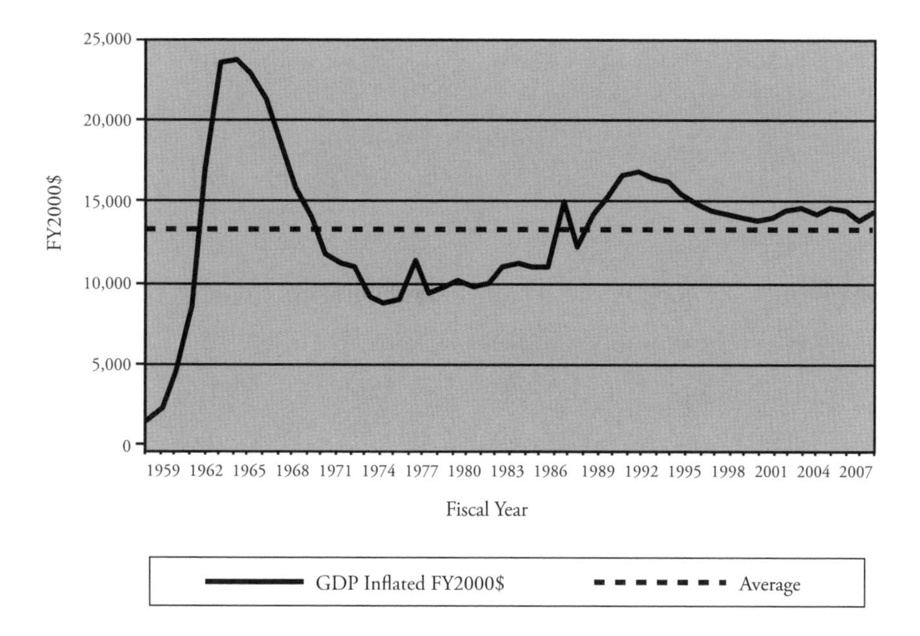

international and commercial partners to be part of turning these ideas into reality. We have a lot of work ahead of us.

While most of these projects and programs will not begin in earnest until after the shuttle is retired, we need to plan now for the work ahead. I want the young people in NASA and at our contractors who are working today on the space shuttle, space station and exploration programs to know that they will have more to do beyond the 5-year government budget horizon; and I want even younger people in colleges and universities to know what the future holds for them if they join us in this journey. I also want to make it perfectly clear to our current workforce that, if funding for NASA remains as we anticipate, I do not foresee any need for reductions in force at NASA. I realize that this has been the subject of considerable unpleasant speculation over the past several years, primarily due to the lack of stability in our programs as we struggled to pay for space shuttle return-to-flight costs and numerous other reductions and redirections in NASA's budget. My hope is that those days of instability are behind us.

Figure 2. GDP Inflated NASA Budget, Historical and Projected

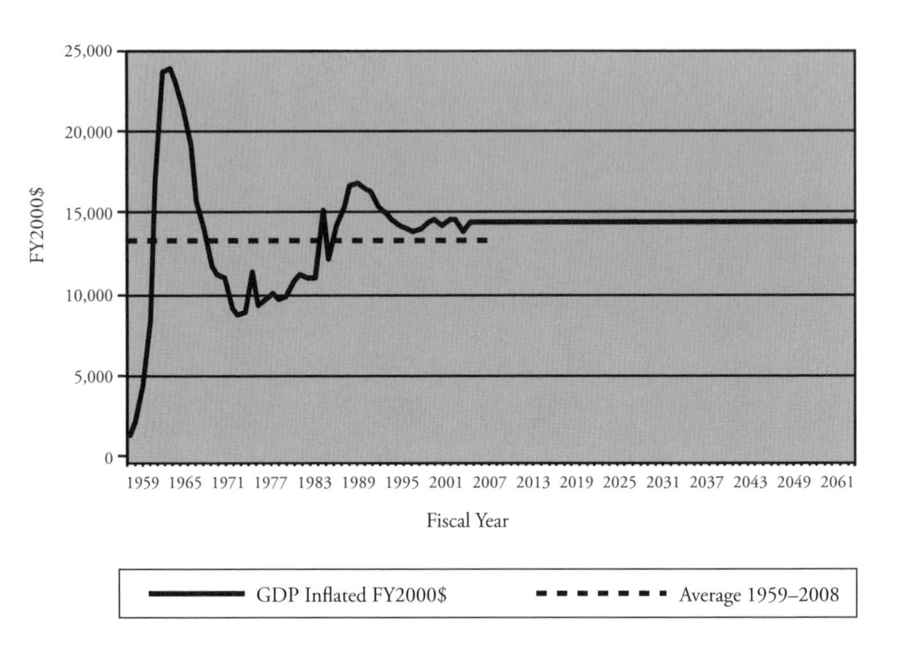

NASA's most important resource—even beyond the budget Congress provides—is its people, scientists, engineers and technicians who make our nation's space program possible, whether they work directly for NASA or not. And our biggest threat today, our biggest management challenge, concerns those people. Space shuttle retirement and transition to a new system is an upheaval that occurs not even as often as once in a generation. We are trying to manage this challenge as best we can with the resources we have been given. We are all concerned that we not repeat the mistakes of the 1970s in the transition from the Apollo to shuttle programs. What was planned as a 2-year gap in human spaceflight turned into 6. Because we did not sustain the commitment (which had been made in the 1960s) because real-dollar funding in the post-Apollo years was so severely reduced, there was an exodus of talent from the program, squandering national capability and weakening the industrial base. We must not allow that to happen again. We must commit ourselves, as a nation, to the space enterprise. We must commit to leadership, not followership, in space exploration. It matters.

Figure 3. GDP Inflated NASA Funding Aggregated by Decade

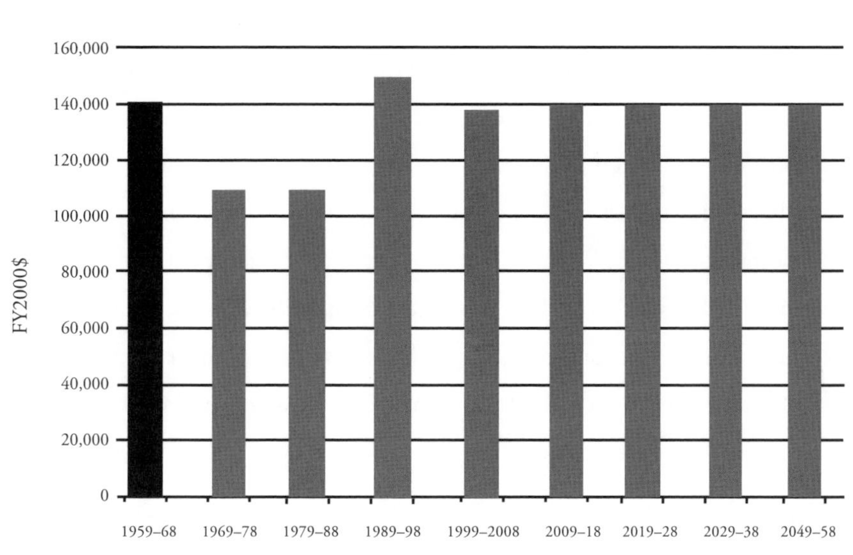

Figure 4. GDP Inflated NASA Funding Aggregated by 15-Year Interval

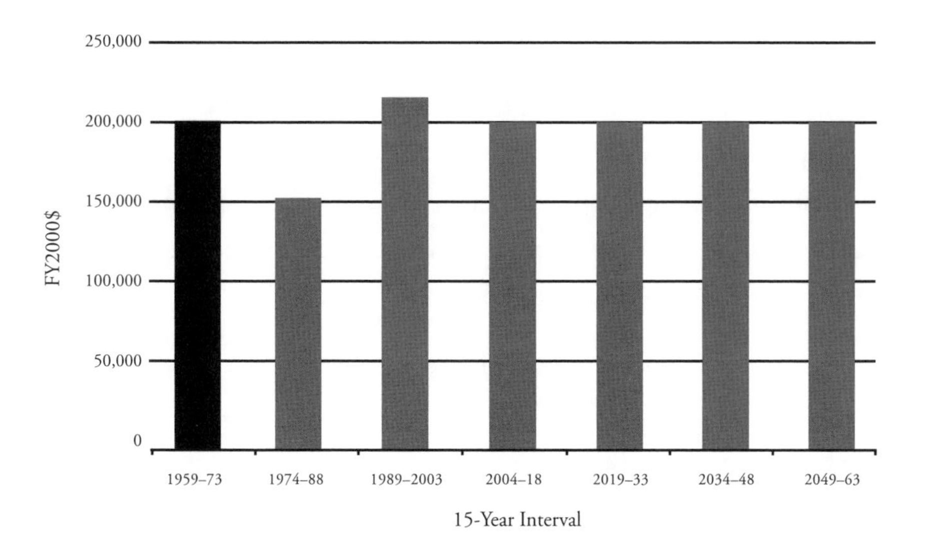

I've touched on some of the rational and acceptable reasons for which we explore space, but I hope you will allow me a moment to share with you a recent lesson that reflects the real reasons why I think most of you in this room decided to go into the space business. Last July, Space Shuttle *Atlantis* was being ferried across the United States after landing at Edwards Air Force Base. Due to weather conditions, the B-747 ferry crew needed to make an unscheduled stop in Amarillo, Texas. Within a few hours, the word had spread on radio and television that the space shuttle had landed in Amarillo. What began with a few passers-by near the airport turned quickly into an impromptu pilgrimage by several hundred and then several thousand people, parents and children to see our nation's space shuttle on display.

Especially poignant in this story is the fact that this airport was named after Rick Husband, Commander of Space Shuttle *Columbia* on February 1, 2003. Those of us in the space business will never, must never, forget the lessons of that day, just as we remember those we learned on January 27, 1967 and January 28, 1986.

I want to conclude on that note of remembrance. We must remember how we felt in October 1957; we must remember how we felt in February 2003; and we must understand how such events change our lives and the course of human affairs, especially as we consider what we choose to do in the future. And for those in the space policy arena who wonder sometimes what the meaning and value of the space program is to the American people, remember the story of that stopover in Amarillo.

We are only 50 years into the development of space, with milestones of both joy and tragedy behind us. Preparing for and having a part in it has filled my life. I have been very lucky.

The Reality of Tomorrow

Michael D. Griffin
Administrator
National Aeronautics and Space Administration
American Astronautical Society
Goddard Symposium
March 5, 2008

The first Goddard Symposium was organized in 1961; and the theme for that year's meeting was "The Interaction of Space Vehicles with an Ionized Atmosphere." This year's theme, "Exploration to Commercialization: Going to Work in Space," deals with some of the many aspects of spaceflight that are not simply about scientific discovery or about the unique engineering problems of operating in this still-new medium. And if you look at the changing themes of this symposium over time, you will notice that this is not an isolated example. The themes of this conference have gradually evolved from strictly technical subjects to the broader implications of spaceflight for human society.

This is good, and it should continue. The state of the art in engineering is a transitory thing at best. Those, like me, who once studied vacuum tube design and programmed in assembly language on "advanced" computers with 16 kilobytes of 16-bit core memory, know this all too well. And the burning scientific questions of one generation are the received wisdom of the ones after that. It seems quaint, and more, to realize today that Einstein received the 1921 Nobel Prize for his elucidation of the photoelectric effect in part because relativity theory was then still too controversial. Our scientific and technical frontiers are transitory, but the deeper questions that devolve, in one form or another, to "What does it all mean?" will always be relevant.

The poster advertising this year's conference shows at first glance the range of space endeavors today. Near Earth, we have burgeoning commercial space tourism highlighting the role of prizes in spurring technological innovation, as with Burt Rutan's *SpaceShipOne*. Moving outward, we will place NASA's Orion Crew Exploration Vehicle in orbit around the moon in the next decade, breaking out of low Earth orbit for the first time in 45 years. And there is a stunning picture of a spiral galaxy bringing to mind the vastness of space with all its possibilities. Of course, this poster is hardly drawn to scale. ... But it is always easier to be an art critic than an artist, so I better not stray too far out of my lane.

However, I do have one serious point to make about how we in the space community try to talk about these larger issues in an understandable way. My hope, as an engineer, is that we do not downplay the technical difficulties of spaceflight to the general public or to our stakeholders in the White House and Congress who don't do what we do for a living. I mean, this is rocket science.

As just one example, the energy which was harnessed to launch *SpaceShipOne* on its suborbital flights is about 2 percent of what is needed to get into low Earth orbit, never mind carry out missions to the moon, Mars and other planets. As I have noted on several occasions, I admire Burt Rutan enormously, and even more so the achievement of him and his team with *SpaceShipOne*. But their achievement was one of breaking (thank heavens) organizational and institutional paradigms not of pioneering new technical frontiers. We in the space community need to communicate these differences clearly.

While I am speaking of commercial suborbital spaceflight, I will note that I also very much hope that NASA researchers and astronauts will be proactive in taking advantage of such capabilities as they are developed by the nation's entrepreneurs. Last week, NASA's Science Mission Directorate issued a request for information on potential human-tended, government-sponsored flight experiments which could be flown on such commercial suborbital vehicles. I hope this request generates a lot of good ideas because all of us at NASA want to figure out how to engage the emerging commercial space sector to advance NASA's goals.

We would not be where we are today if it weren't for doers with big dreams, people like Robert Goddard and Burt Rutan. Burt continues to be an inspiration and mentor for many aerospace engineers, with his offices at Scaled Composites lined with *Aviation Week & Space Technology* magazine covers depicting his varied handiwork along with a Collier Trophy or two on the bookshelf. In many respects, Burt is carrying on the tradition of great engineers whose work was initially dismissed but recognized later.

Robert Goddard, the practical physicist whom we honor with this symposium, was another pioneer whose ideas were initially panned. One of the oft-told tales in the space business is of how, in 1920, *The New York Times* editorial board questioned Goddard's technical acumen, noting that rockets could not operate in the vacuum of space and saying of Goddard: "He only seems to lack the knowledge ladled out daily in high schools." Goddard's simple yet profound response: "Every vision is a joke until the first man accomplishes it; once realized, it becomes commonplace."

Of course, the truth eventually surfaced. Forty-nine years later (as Apollo 11 was on its way to the moon) and buried on Page 43 of its July 17, 1969 edition, *The New York Times* grudgingly admitted: "It is now definitely established that a rocket can function in a vacuum as well as in the atmosphere. The Times regrets the error."

But the quote from Goddard that I like the most is this one: "It is difficult to say what is impossible, for the dream of yesterday is the hope of today and the reality of tomorrow," because it is my own hopes for the reality of tomorrow that I would like to discuss with you this morning. Unfortunately, it is in my nature to be a realist. Some in the space community wish that I were more a cheerleader or a font of inspiration, but those of you who have known me for decades know that such irrational exuberance is not in my character. Thus, I sometimes point out the harsh realities with pernicious facts about our nation's space and aeronautics enterprise. Just the facts.

For example, it is a fact that as a matter of national policy spanning multiple administrations and congresses, NASA simply is not allocated the budget resources to accomplish all of the many and varied space and aeronautics missions that our many constituencies would like us to do. The president's request for NASA in fiscal year (FY) 2009 is $17.6 billion out of $3.1 trillion for all U.S. government spending, less than 0.6 percent of the entire federal budget. During the development peak of the Apollo program, NASA received 4.4 percent of the federal budget and employed over 400,000 civil servants and contractors across the country; today, we employ maybe 90,000 people. Adjusted for inflation, NASA's budget is $3 billion less per year than it was the last time I was at NASA in the early 1990s. That is about 20 percent less buying power today than when I was associate administrator for exploration. These are just facts.

To the Bush Administration's strong credit, NASA's budget has kept pace with inflation at a time when most other non-defense domestic discretionary budgets have not, showing the priority given to the agency's mission after the prior years of decline. In keeping with my sardonic side: when you care enough to send the very best, send money. This administration has. But the reductions made in prior years have not been reversed. Just a fact.

During this period of declining NASA budgets since I left the agency in 1994, no viable replacement for the space shuttle was developed; aeronautics research funding declined significantly; funding for new space technology dried up; and science funding rose to an all-time high percentage of the agency's budget. Just facts.

Thus, a major focus during my tenure as administrator has been to find some semblance of budgetary balance among the competing priorities of NASA's overall budget within the resources provided, while also returning the space shuttle to flight after the *Columbia* accident, completing the assembly of the International Space Station, and building new crew launch systems.

As Admiral Hal Gehman noted in the report of the Space Shuttle *Columbia* Accident Investigation Board 5 years ago, "previous attempts to develop a replacement vehicle for the aging shuttle represent a failure of national leadership." Because of that, we now face a gap in U.S. human spaceflight capabilities after shuttle retirement and are reliant upon the Russian *Soyuz* system for transport to and from the space station that we have built.

Thus, I must admit that our post-Apollo budget history and our posture today color my view of the reality of tomorrow. While this is hardly inspirational rhetoric, my hope for today to become the reality of tomorrow is that NASA's budget will continue to grow at least in accord with the Office of Management and Budget's (OMB) inflation index for the next 50 years. This kind of budget, and this kind of stability, will produce a pretty good outcome if we manage it with a proper sense of purpose.

About 1 year ago I wrote a lengthy article for *Aviation Week & Space Technology* concerning what we can realistically afford to do under such a budget scenario. If we make the necessary strategic investments and maintain the sense of purpose that I find around the agency today, then we can indeed be back on the moon by 2020, have a lunar base by mid-decade and be on Mars by the mid-2030s.

After I wrote the article, I was criticized for painting a rosy picture by using the prescribed OMB inflationary index when some believe that, through no fault of our own, the real costs of aerospace goods is much higher. Others believe that if aerospace costs are rising disproportionately in comparison to other high-tech sectors, the fault lies within our community and our culture and should be addressed there. This is not an esoteric argument but one that has profound implications for NASA's real purchasing power and ability to plan for multiyear projects and programs in an era when the costs of raw materials and high-tech labor are increasing higher than inflation. As Albert Einstein observed, "The most powerful force in the universe is compound interest." Thus, all I can say is that until someone else demonstrates a better model for inflation than OMB's prescribed index, while I might not be full of irrational exuberance, I am cautiously optimistic in my hopes for the reality of tomorrow.

So, given what we know of NASA's budget over the past several decades and what we might reasonably project for the future, we can either bemoan the underfunding of our nation's efforts in space and aeronautics research, wallow in pity about the lack of progress being made or find some more productive and constructive approach to our problems. I choose the latter. So let me talk now about my second hope for the reality of tomorrow.

Let us speak openly and honestly about the problems we face in carrying out our nation's space program. Over the course of my career in this business, I have often been disheartened by the large number of diverse "entrepreneurs" in search of NASA funding who place their self interests over the greater good of the aerospace community. They do not respect the priorities set out for NASA by our duly-elected stakeholders in the White House and Congress or even the priorities of their own respective science communities in decadal surveys by the National Academy of Sciences. Even worse, the rift and harsh rhetoric between proponents of robotic science and human spaceflight does not help our nation's overall space effort one iota, but it does cause division that weakens us. If we wish a better reality for tomorrow, we as a community must police this behavior; those who engage in it must be made to feel, and be, unwelcome in the community at large. My hope for today is that there will in the future be more respect for each others' work.

I must also point out that there have been many instances where proponents of individual missions have downplayed the technical difficulty and risk of their individual mission or grossly underestimated the cost and effort involved to solve the problems in order to gain "new start" funds for a particular project. Everyone knows that, once started, any given mission is nearly impossible to cancel; so the goal becomes that of getting started no matter what has to be said or done to accomplish it. I am speaking here not only to industry and scientific investigators but also to organizations within NASA. This is a matter of integrity for our community. NASA managers, the White House and Congress have seen this behavior too many times, and the agency has lost a great deal of credibility over the decades as a result. There was a time—I remember it, and many of you will also—when what NASA said could be taken to the bank. Anyone here think it's like that today? Show of hands? I didn't think so.

I have spent a good portion of my time as administrator trying to rebuild that credibility with more rigorous technical review and independent cost estimating processes. But folks, we are in this together. We will not be trusted with more funding to carry out great, new and exciting space missions in the future, human or robotic, if we oversell and under-deliver on our commitments today. Across the board, we must be realistic in our assessments of cost and technical risk if we are to be trusted with funds provided to us by the American taxpayer.

We must also change some of the ways in which NASA conducts its business. No one who has worked both in government and in the private sector can fail to note the efficiency of commercial operations as compared to those of government. Just as we are conceiving plans for commercial suborbital missions, we also recently awarded a contract to the Zero-G Corporation to use their aircraft for NASA's microgravity experiments and astronaut training. We will be taking a hard look over the coming months to determine whether NASA should continue to own and operate its current C-9 aircraft for parabolic flights. A couple of weeks ago, NASA signed a new, funded Commercial Orbital Transportation Services (COTS) Space Act Agreement with Orbital Sciences and another agreement with SpaceX to spur private industry investment to develop and demonstrate cost-effective cargo and crew transport to the space station. Our goal is for NASA to be able to purchase commercial delivery of goods to the space station after the shuttle is retired in 2010. With the space station as a national laboratory, we are opening it up to commercial use as well as to other agencies of government like the National Institutes of Health. We are going to work in space.

Just down the road from here, the men and women of the Goddard Space Flight Center are going to work in space. The reality of tomorrow is that over the next few years, Goddard is playing a major role in launching nearly a dozen science missions—one of the busiest periods in the center's illustrious history.

First up this year is the Gamma-ray Large Area Space Telescope (GLAST), a marriage of astronomy and particle physics to study black holes and the physics of extreme energies and what composes dark matter. This project, set to launch in late May, is also a marriage between NASA, the Department of Energy and research institutions in France, Germany, Japan, Italy and Sweden. Indeed, over half of NASA's missions involve some form of international cooperation. One month ago, we asked the science community and the public to offer recommendations on renaming the GLAST mission. Let's hear from you.

Perhaps the most inspirational mission NASA hopes to carry out later this summer is the final servicing mission to the Hubble Space Telescope by the crew of STS-125 on Space Shuttle *Atlantis*. The Hubble servicing mission turns the unhealthy schism between human and robotic spaceflight into a meaningless argument. With four previous service calls, our astronauts risked their lives to correct the Hubble's flawed optics, install new gyros, batteries, solar arrays as needed and install a series of powerful new instruments, dramatically boosting its capabilities and performance. It is a marriage of human ingenuity and state-of-the-art scientific know-how and perseverance. I saw that Frank Cepollina is speaking here at the Goddard Symposium tomorrow and I would rather you hear directly from him what we have accomplished with the Hubble, our progress in carrying out this mission and what's next.

The scientists and engineers of Goddard are also completing the development work for the Lunar Reconnaissance Orbiter (LRO), NASA's first mission to our closest celestial neighbor in a long time. The LRO, to launch late this year with the Lunar Crater Observation Sensing Satellite (LCROSS) lunar impactor, will create the most accurate and comprehensive topographic maps of the lunar surface to date, vital for pinpointing landing sites for future manned missions. As Doug Cooke observed last week when unveiling the recent Goldstone radar maps of the moon's south pole around the Aitken Basin, an area of great potential interest as a landing site, "We now know that the south pole has peaks as high as Mt. McKinley and crater floors four times deeper than the Grand Canyon. There are challenges that come with such rugged terrain."

But American explorers, people like Daniel Boone, Meriwether Lewis and William Clark, Zebulon Pike and many others who followed are used to rugged terrain. Thus, when our astronauts return to the moon for the first time in nearly 50 years, they will be blazing new trails and carrying on a rich tradition of exploration on the "New Frontier" as President John F. Kennedy framed our nation's first forays into space.

The LRO mission is one of many demonstrating the marriage between human and robotic space exploration. The LRO, Mars Reconnaissance Orbiter, the Mars Exploration Rovers, the Phoenix Scout and Mars Science Laboratory missions are examples of some of the robotic missions that will enable future manned exploration as we take the first rudimentary steps out into our solar system.

Another mission which will also enable future manned exploration is slated for liftoff at the end of this year. NASA's Solar Dynamics Observatory (SDO), also built here at Goddard, is another mission where the men and women of Goddard are turning into the reality of tomorrow. The SDO will be an unblinking eye on our sun using the technique of helioseismic imaging to look inside our sun and deliver images with 10 times better resolution than even high-definition television. This mission will revolutionize our ability to forecast solar storms that disable satellites, scintillate Global Positioning System (GPS) signals, cause interference to satellite communications and cell phone calls and endanger our astronauts from solar radiation. The importance of space weather data is evident in that more than 250,000 customers from 150 countries around the world receive almost 200 data products monthly, approaching 30 million file transfers a month.

Very recently, the Solar Terrestrial Relations Observatory (STEREO) mission, built by my nearby alma mater the Johns Hopkins University Applied Physics Laboratory, witnessed the tail of a comet being completely blasted away by a coronal mass ejection from the sun. It's an amazing thing to watch; and it's posted on the STEREO Web site if you haven't seen it already.

The Goddard Space Flight Center is also our nation's premier center for global warming research and developing climate change monitoring satellites. NASA satellites, many built by Goddard, supply more global climate change data than those of any other organization in the world. NASA is also the largest contributor to the inter-agency Climate Change Science Program (CCSP), providing the more grant-based funding for such research than any other organization. We can take great pride in the recognition on the part of policymakers and the public of the value of the research coming from NASA's Earth scientists. NASA's work in developing so-called green technologies, like fuel cells in cars and clean water treatment systems for rural villages in Africa and Iraq, is featured in an article "Space is the Place" in this month's issue of *Sustainable Industries*. This is one part of turning the hope of today into the reality of tomorrow.

Today, we are living in exciting and tumultuous times; and now is the time; this is the juncture; and we are the people to turn our Vision for Space Exploration into reality. In order to do so, we must cast aside many old chestnuts—like the divisiveness between those of us who work on robotic space and those who work on human missions—to appreciate that many things we do in space are conducted for entirely other reasons than science.

So in conclusion, I would like to repeat a story I told 2 months ago in Houston, the heart of our nation's human spaceflight endeavors. We lived in similarly tumultuous times 40 years ago in 1968 when I was a college sophomore. Former NBC News Anchor Tom Brokaw recently wrote a book titled *Boom! Voices of the Sixties* about that year. With the Vietnam War, the assassinations of Martin Luther King and Robert Kennedy, the struggle for civil rights and women's rights, protests on college campuses and the presidential campaigns and election that year, 1968 was a time of great upheaval for our nation.

Tom Brokaw points out a number of parallels between 1968 and today. He ends his analysis of that year with a reminder of what the inspirational Apollo 8 mission in December 1968 meant to our nation and the world. On Christmas Eve, the crew of Apollo 8—Frank Borman, Jim Lovell and Bill Anders—read from the book of Genesis as they cruised in orbit around our moon; they saw our fragile Earth rise over the barren horizon of the moon; they took the first photograph of our Earth in full with blue oceans, white clouds, green and brown land without any artificial national borders. Jim Lovell looked back at Earth, held up his thumb and blocked Earth out from his view. He has since said that he realized in that moment how small the world he once knew was when compared to the vast frontier of space. With all the turmoil of 1968, the Apollo 8 mission and this transcendent moment helped all of us to realize that we must overcome our common struggles if we are to achieve better things for ourselves and future generations.

With both human and robotic eyes, we see our planet, our solar system and our universe in ways we never imagined. Space exploration has brought many nations together in ways unimagined when NASA was first founded 50 years ago. "It is difficult to say what is impossible, for the dream of yesterday is the hope of today and the reality of tomorrow." Robert Goddard speaks to us even today.

Human Space Exploration: The Next 50 Years

Michael D. Griffin
Administrator
National Aeronautics and Space Administration
(This article first appeared in the March 14, 2007, issue of *Aviation Week & Space Technology*.)

"Prediction is difficult, especially the future," said Quantum Physicist Niels Bohr, and no one has since captured the underlying concept quite so cleverly. But having been foolish enough to accept the challenge of speculating upon where the next 50 years will take us in human space exploration, the first question to be answered is: where to begin? What is the global view that can best shape our thinking? It is so very easy to be completely wrong since a variety of radically different futures in spaceflight can be presumed with equal apparent credibility today.

For example, it might be that after completing the construction of the International Space Station and retiring the shuttle, the excitement inherent in a new reach outward from low Earth orbit will appeal to the next generation, leading to a vigorous, technology-driven program; a plan to reach the surface of Mars by the late 2020s; and the will to sustain and build upon that early presence. Or, interest in human space exploration could once again be motivated by competition among spacefaring nations, leading to a modern version of the Space Race of the 1960s, producing substantial progress but for reasons unsustainable in the longer term. It could be that the unchecked growth of entitlements, the generational commitment of resources necessary to combat terrorism and a continued downward trend of interest by American students in mathematics, science and engineering education will combine to make the civil space program as we have known it irrelevant to the lives of our grandchildren's generation. Or the truth could lie in some other direction entirely; maybe human spaceflight in the next 50 years will be dominated by tourism, first suborbital then orbital, with relatively little in the way of independent government activity.

The one thing of which we can be certain is that in trying to envision the world of 2057, two generations in the future, we will be wrong. We will be wrong in our assumptions about the larger context of world culture and civilization in which space exploration exists; and we will be wrong even in the narrower context that is the subject of our attention here. Even the most cursory review of some of the key events in the development of spaceflight shows the need for great humility by anyone writing an essay on the likely state of space exploration in 2057.

For example, who would have supposed in early 1957 that the Soviet Union, and not the United States, would loft the first artificial satellite into Earth orbit, the first robotic lunar probe and the first man into space? And who would ever have predicted that the United States, stung by losses in a competition in which it had not even known it was engaged, would or even could respond by carrying out the first lunar landing 8 years and 2 months after declaring the goal? Most then-knowledgeable observers believed that such a feat was unlikely to be achieved much before the end of the 20th century, if then. Not even the most visionary of hard science fiction authors—Asimov, Clarke, Heinlein—imagined that it could occur as early as 1969. And then having spent $21 billion (in mid-1960s dollars) to develop the transportation system to make such a thing possible, was it even conceivable that such hard-won capability would be utterly discarded within a few years? Who would have imagined it? And yet it happened.

With those thoughts in mind to encourage an appropriate humility, it is nonetheless natural to wonder how we might develop a vision of the future that is the least likely to be terribly wrong. How can we extrapolate today's world in such a way as to avoid the most outrageously wrong predictions?

Most of the present discussion will focus on the U.S. government civil space program. I will have some comments on the international scene and on the possible role of commercial space; but for much of the next five decades, the U.S. government will be the dominant entity in determining the course of human space exploration. We will, I hope, develop robust international partnerships that will enormously enhance the value of space exploration. And we must do everything possible to provide an accepting environment for commercial space entities, standing down government capability in favor of commercial suppliers whenever it becomes possible to do so. But with that said, the U.S. today is spending more than twice as much on civil space per capita, as any other nation, and I believe this situation is unlikely to change significantly for some time. Commercial space firms offer great promise but, so far, limited performance. For a while yet, it is the U.S. government, through NASA, that determines the main course of human spaceflight.

Of course, manned spaceflight is broader than exploration, and over the next decades it is to be expected that other entities besides the U.S. government, both commercial and international, will be conducting human spaceflight activities. A spacefaring civilization cannot be the realm only of government employees and government sponsored engineers and scientists, though a bias toward such groups is clearly one attribute of a frontier activity. But if we understand that broader participation is desirable, U.S. human space exploration programs can be conducted so as to encourage, rather than minimize, such. Doing so will, in my opinion, be a key to its survival and prosperity—a point that I will make again in what follows.

But let us now focus our attention on more specific matters. The most straightforward extrapolation is to assume that the future will, on average, be much like the past in regard to key assumptions. Since no aspect of government civil spaceflight is more crucial than the funding allocated to it, let's consider NASA's funding history for the last 50 years and try to make a reasonable yet conservative projection as to what we might receive in the next 50. And then let's consider what that funding might allow us to do, setting aside unforeseeable political upheavals. To understand where we might go, we must understand where we have been. And I think we need a better understanding of our history than is commonly the case.

Any assessment of historical or projected budgets necessarily must be done in constant, inflation-adjusted dollars. This fact leads inevitably to the question of what inflation index should be used because long-term assessments are sensitive to that choice. Many choices are possible. The Bureau of Labor Statistics maintains the familiar Consumer Price Index (CPI), applicable to the U.S. economy at large, that is, the Gross Domestic Product (GDP). However, the CPI is not the best measure of inflation for government spending primarily because the "market basket" of goods and services applicable to the private and public sectors of the economy are very different. The best use of the CPI in connection with government programs is in the estimation of the constant-dollar "opportunity cost" of government activities to citizens. Government services are purchased by taxpayers with CPI-adjusted tax dollars; money paid in taxes is money not available to consumers to purchase other goods and services.

The Office of Management and Budget (OMB) publishes several inflation indices applicable to different portions of the government sector. For government R&D activities, including those at NASA, the OMB prescribes the use of the so-called "GDP (chained) Price Index" (*www.whitehouse.gov/omb/budget/fy2008/sheets/hist10z1.xls*). Without delving into the merits and shortcomings of various indices, our discussion of inflation-adjusted NASA funding will employ this index. While fiscal analysis across several decades is sensitive to the choice of inflation index, the present discussion is not significantly influenced by the choice of the GDP chained index versus other OMB indices. Unless specifically stated otherwise, all fiscal discussions in this essay are couched in terms of fiscal 2000 dollars with inflation adjustments according to the OMB GDP (chained) Price Index.

Figure 1 shows the constant-dollar budget for NASA's first 50 years, 1959–2008, in fiscal 2000 dollars and includes the assumption that the agency will be funded in fiscal 2008 at the level of the president's request. Data for other fiscal years is historical. The anomalous funding bump in fiscal 1977 is due to the inclusion of a fifth "transition quarter" in that year since in 1976 the fiscal year boundary was shifted from July 1 to October 1 where it remains today. Major events in NASA's history—the "Apollo Peak," the post-Apollo aerospace depression and the supplemental funding provided by Congress in response to the *Challenger* disaster are all clearly visible in figure 1.

As seen, NASA today is funded at a constant-dollar level slightly higher than the agency's historical average. With proposed growth in the president's budget for fiscal 2008–12 roughly matching the anticipated rate of inflation over the next several years, agency funding is expected to remain slightly above the 50-year average.

In an attempt to offer a reasonable but conservative vision for government civil space activities, let us assume that NASA continues in fiscal 2013 and beyond to be funded in constant dollars at the average level of the president's request for fiscal 2008–12. This is illustrated in figure 2 with the average out-year budget assumed to be $14.2 billion in fiscal 2000 dollars. We in the space community will certainly hope for more but we should not expect less. More properly, we should expect to perform in such a manner—actually delivering a bold, exciting, efficient and effective space program, instead of PowerPoint charts with hopes and dreams—that policymakers do not want to provide less!

The year-to-year budget profile will show some variability, of course, but we should expect considerably more strategic and fiscal stability than was evidenced in the agency's first few decades. Minor annual variations should not affect the larger picture; on the 5- to 15-year cycle of developmental space programs and projects, it is the average level of funding which is the most significant parameter. The total funding received by the agency over a significant period, a decade or more, together with stability of strategic goals, largely determines what can be accomplished.

Figure 3 offers a different view of historical and projected NASA funding for the past and future 50 years. Funding is aggregated by decade and incorporates the assumption of a stable constant-dollar budget embodied in figure 2. Figure 4 provides a similar view with funding aggregated in 15-year intervals and constant inflation-adjusted funding assumed through 2063. This 15-year assessment period is particularly convenient since essentially all Mercury, Gemini, Apollo and Skylab development and operations are captured within the first 15 years of NASA's history.

Figures 3 and 4 offer what might be a new perspective for many. From a decadal viewpoint, the "Apollo peak" in NASA funding, regarded by so many as the agency's halcyon period, is a myth. In truth, NASA received funding well above its historical average level for only 5 years, 1964–68, followed by a lengthy and debilitating reduction. But when averaged over decadal or 15-year time scales, the nation's civil space program has experienced no particularly noteworthy funding peaks. The highest historical funding period was actually in the decade (or 15-year interval) centered on the early 1990s and not during Apollo. Further, if we assume funding stability in constant dollars (shown in figure 2), the total in every subsequent decade will match that of the Apollo development decade of 1959–68. Expressed in a slightly different way, NASA could carry out a complete Apollo-scale effort every 15 years between the present day and the 100th anniversary of *Sputnik*.

Let us now address another time-honored belief about the Apollo era. When we talk about an "Apollo-scale effort," it is important to understand that, contrary to conventional wisdom, we are not talking about an agency devoted exclusively to human exploration. The funding record clearly shows that the "Apollo era" was actually quite a lot more than just that.

In the Apollo development decade of 1959–68, human spaceflight received 63 percent of the budget. Funding specifically for Apollo from its inception in fiscal 1961 to its completion in fiscal 1973 was about $105 billion in fiscal 2000 dollars. If Mercury ($1.9 billion), Gemini ($5.1 billion) and Skylab ($12 billion) are included, the entire human spaceflight program from 1959–73 received about $125 billion, or 61 percent of the $206 billion allocated to NASA during this period. Little has changed in this regard; today, the president's fiscal 2008 budget request assigns 62 percent of NASA's funding to human spaceflight.

The list of achievements in both aeronautics and space science from 1959–73 is long and impressive. Aeronautical accomplishments of this era include 199 research flights of the three X-15 rocket planes; the development and flight testing of a half-dozen lifting-body designs; groundbreaking work in computational fluid dynamics; development of the supercritical wing and the digital fly-by-wire flight control system; and (in conjunction with the Air Force) major roles in the XB-70 and YF12A programs. The "Apollo era" was a true golden age for aeronautics research, which was allocated 6 percent of the NASA budget from 1959–68.

In space science the list of accomplishments is, if anything, even more impressive. The "Apollo era" saw dozens of Explorer missions including the Radio Astronomy Explorer and Atmospheric Explorer series; a dozen Pioneer missions including Pioneers 10 and 11 to Jupiter and Saturn; Rangers 1–9; Surveyors 1–7; Mariners 1–10; the Orbiting Solar Observatory, Orbiting Geophysical Observatory and Orbiting Astronomical Observatory series; as well as most of the money for two Viking missions to Mars launched in 1975. The Television Infrared Observation Satellite (TIROS), Nimbus and Environmental Science Service Administration (ESSA) series pioneered the development of weather satellites. The "Apollo era" was also a golden age for space science, which received 17 percent of the NASA budget from 1959–68.

About 10 percent of the 1959–68 budget was devoted to space technology development, including space communications technology, and 4 percent was devoted to "Other", that is, university support and cross-agency activities.

The summary below shows a "then and now" comparison. In contrast to oft-repeated claims, human spaceflight is not growing relative to other portions of the NASA portfolio and is not "eating everyone's lunch."

Category	1959–68	FY08 Request
Human Space Flight	63%	62%
Science	17%	32%
Aeronautics	6%	3%
Comm. & Space Tech.	10%	0%
Cross-Agency Support	4%	3%

The historical record provides clear evidence that it is possible to have robust, co-existing programs of human exploration, space science, aeronautics and technology development in a single agency funded at a level essentially the same as we presently receive. So what might the future offer?

Let us assume for the present discussion that, over the long term, manned spaceflight will continue to receive 62 percent of the NASA budget. Again assuming inflation-adjusted funding at $14.2 billion per year on average, it follows that human spaceflight will be allocated $8.8 billion annually or $132 billion in each 15-year period in fiscal 2000 dollars.

Next, we must recognize that "the future" really does not and cannot start until after 2010. Until then, we are engaged in completing a long-standing commitment to the space station with no other option besides the space shuttle to do it. At present funding levels, we cannot afford to develop new human spaceflight systems without the money that becomes available following shuttle retirement.

Despite the concerns of those—emphatically including myself—who worry about the gap in human spaceflight between the retirement of the space shuttle and the availability of the new Constellation systems, Orion and Ares, we must stay on our present course and retire the shuttle in 2010 if there is to be a future for human spaceflight. The shuttle offers truly stunning capability greater than anything we will see for a long time, but the expense of owning and operating it or any similar system is simply too great. Any new system, to be successful, must offer a much, much lower fixed cost of ownership. The space shuttle was designed to be cost effective at a weekly flight rate, a goal that was never credible if for no reason other than the fact that the funding for so many payloads to fly on it was never remotely available. And if there were a predictable requirement for 50–60 government-sponsored payloads to be flown annually, that fact should be treated as a market opportunity for a private, not government, space transportation enterprise. A government human spaceflight system must be designed to be cost effective at the half-dozen or so flights per year that we can expect to fly.

But if the bad news is that "the future" doesn't start until after 2010, the good news is that it is only 4 years away. And in the 45 years thereafter by the centennial anniversary of *Sputnik*, we can expect to receive at least as much money as was necessary for Apollo three times over. And despite the limited funding for exploration in today's NASA budget, we will have a bit of a head start because we're making considerable progress toward the deployment of Orion and Ares even while flying out the shuttle/space station manifest. So what will we do with this money?

Most of the next 15 years will be spent recreating capabilities we once had, and discarded. The next lunar transportation system will offer somewhat more capability than Apollo. It will carry four people to the lunar surface instead of two for a minimum duration of 1 week rather than a maximum duration of 3 days. But in all fairness, the capabilities inherent in Orion, Ares I and Ares V are not qualitatively different than those of Apollo and certainly are not beyond the evolutionary capability of Apollo-era systems had we taken that course. But we did not and the path back out into the solar systems begins, inevitably, with a lengthy effort to develop systems comparable to those we once owned. It will cost us about $85 billion in fiscal 2000 currency to get to the seventh lunar landing by 2020.

The above assessment is, for many, a bitter pill to swallow. Not only is it depressing for advocates of human exploration to face the fact that so many years will be spent plowing old ground but there is also the question of why it will take so long. Again, the answer is captured in the funding profile. We are indeed receiving today in any given 15-year period, the same real-dollar funding as in the 15 years of the Apollo era, but we are not receiving it on the same schedule. The brief, enormous funding peak of the mid-1960s allowed the Apollo systems to be developed and procured in parallel. Today's systems must be developed serially. And that is why the job will not be done, this time, in 8 years. But that is also why we will not incur the disastrous divestiture of talent and technology that occurred in the 15 years after Apollo between the early 1970s and the late 1980s.

In the long run, to return to the moon or go to Mars and beyond, stability is to be valued more than going in the shortest possible time. As we move forward into our next 50 years, this must be fully understood by both policymakers and the public or we will forever be answering the question as to why we work so slowly compared to the Apollo generation. Civil space exploration beyond low Earth orbit must have the stability in strategy and funding that was lacking the first time around. This will only be provided by policymakers if a clear link is established between predictable results and predictable purpose, strategy and funding. I believe we will succeed in forging this new paradigm—the opposite of the Apollo "man, moon, decade" paradigm—but we must devote considerable attention to doing so.

What will be done with the lunar transportation capability that is being developed? By 2020 we will have this capability and with it choices to make. We can choose between a lunar program devoted to sortie missions or one devoted to building up a lunar outpost. And we can choose between the level of effort we intend to focus on lunar activities versus initiating development for Mars missions. In company with other space agencies around the world, we at NASA have focused on an outpost-centered lunar exploration strategy. I believe this will be preferred over a sortie-only strategy for the reasons that it provides a much more effective avenue for international partnership and because it provides the greatest opportunity to learn on the moon what we need to know to go to Mars. But of course nothing prevents a sortie mission to any location on the moon that is of sufficient interest to justify the expenditure of funds. So again, let us look at what is fiscally possible.

It is to be hoped and, I believe, expected that the next era of space exploration will be international in scope in much the same fashion as the development of the space station today. Whatever might be said of the space station program—and there cannot be much that has been left unsaid—it has pioneered a path to the development of a major international space facility. There are lessons learned in so doing that we will take with us out into the solar system. These lessons will be the most enduring and ultimately most valuable contributions the space station can make. We will be applying them on Mars 50 years from now.

The United States is developing the transportation system which will allow access to the lunar surface for the first time in a half-century. This is the highest "barrier to entry" for exploration beyond low Earth orbit, one which essentially exhausts the contribution that we can make to a lunar outpost in the next 15 years. If there is to be a lunar presence significantly beyond merely getting there and getting back, if there is to be a human tended outpost, much of the early capability must be developed by international partners. But outpost sustainability, at least in the early years, will largely depend upon Orion and Ares.

I believe that by 2021–22 we will have regained enough experience in lunar spaceflight operations that we will be able to undertake a modest but sustained and sustainable program of lunar outpost development and utilization. I will also venture to say that by 2022 the space station will be definitely behind us. We will have learned from it what we can, but there will come a time when the value of the work being done onboard the facility will be judged not to be worth the cost of sustaining its aging systems and it will be brought down. I don't know when this will occur and I am not sure it is predictable other than in a statistical sense, but I believe that by 2022 or thereabouts it will have happened. And when it does, the resources that have been used for space station support can be applied to the support of a lunar outpost.

For the sake of argument and nothing more, let us say that in 2022 we will begin a sustained lunar program of exploration and development consisting of three manned missions (two outpost crew rotations and one sortie) and one unmanned cargo mission per year utilizing three Orion and Ares I vehicles and four Ares V launches. Present projections assume a cargo capacity of 6 metric tons on a lander carrying four crewmembers and 20 metric tons on a cargo lander at a marginal cost of about $750 million for a human mission and $525 million for a cargo mission. The marginal cost in fiscal 2000 dollars for this nominal lunar program will thus be about $3 billion.

These marginal costs do not include an allocation of the fixed costs of production and operations which will be assigned to each flight. Let us assume a fixed-cost support base of $1 billion annually with about a third of that for the shuttle today, equivalent to roughly 6,000 full-time employees at average fiscal 2000 labor rates. We should all work to make it much less; but this is an appropriately conservative estimate for the present. This yields a sustained lunar program costing no more than $4 billion per year, leaving $4.8 billion annually in the human spaceflight account to be applied to new development priorities.

By the 2020s, we will be well positioned to begin the Mars effort in earnest. The lunar campaign will have stabilized; a human-tended outpost will be well established; we will have extensive long-duration space experience in both zero- and low-gravity conditions, and it will be time to bundle these lessons and move on to Mars, which does not imply that we will bring lunar activities to an end. Quite the contrary. My prediction is that the moon will prove to be far more interesting, and far more relevant to human affairs, than many today are prepared to believe. But by the early 2020s, it will be time to assign a stable level of support for lunar activities and set out for Mars.

The development of the Orion, Ares I and Ares V transportation system is being done in a way that provides a substantial capability for subsequent Mars expeditions. In particular, we expect the Orion crew vehicle (or a modest upgrade of it) to provide the primary transportation from Earth to whatever transportation node is used for the assembly of the Mars ship and to be the reentry vehicle in which the crew returns home at the end of the voyage. The Ares V cargo vehicle will provide, with no more than a half-dozen launches, the 500 metric tons or so which is thought to be necessary for a Mars mission based on present-day studies. As a perspective on scale, this mass is about 25 percent greater than that of the completed space station.

It is difficult to estimate the non-recurring cost of developing a Mars mission that is initiated some 20 or more years in the future and especially so when a specific mission architecture has not yet been formulated. But reasoned estimates can be made. A small group co-chaired by *Skylab* and shuttle astronaut Owen Garriott and me made an attempt to do so in a study conducted for The Planetary Society in 2004. While necessarily omitting many important details, a reasonable approach based on mission mass, consistent with modern cost estimation algorithms, was outlined. It was concluded that, following a decadal hardware development cycle, nine Mars missions could be conducted over a 20-year period for a total cost of approximately $120 billion in fiscal 2000 dollars or $6 billion per year, significantly less than we are spending on the shuttle an space station today. (If this seems low, it should be noted that the development cost of the heavy-lift transportation system is allocated to the earlier lunar program. The Mars program would pay only the marginal cost of transportation.)

Allocating an across-the-board 30 percent reserve at this stage puts the cost of a 30-year Mars exploration program at $156 billion in fiscal 2000 dollars. Of this, approximately $70 billion consists of development cost with reserve. If $4.8 billion per year is available in the human spaceflight account, then the Mars mission development cycle will require about 15 years. Thus, if we begin development work in 2021, we will be able to touch down on the Martian surface in about 2037 with follow-on missions every 26 months thereafter for the next two decades.

So there we have it, at least for the U.S. civil space program. At present levels of real-dollar funding, by 2057 we can celebrate the 35th anniversary of a lunar base, which will be growing in capability at the rate of 30 metric tons per year, even without assuming any international partner contribution to logistics (which I believe is overly conservative).

We can celebrate the 100th *Sputnik* anniversary in conjunction with the 20th anniversary of the first human Mars landing. And we can do all of these things even with what I would consider the pessimistic assumption that we receive no more money in constant dollars than we do today. Indeed, there should be money available for missions to interesting near-Earth objects, a separate challenge which we will come to understand offers huge opportunities for those seeking to develop a spacefaring civilization.

That's what I see ahead for the American space program. What about the rest of the world? Both Russia and China have domestic human spaceflight capability today; indeed, the space station program would be in very difficult straits without Russian crew and cargo services. Other nations or alliances—Europe, Japan, India, Brazil and others—could develop similar capabilities within a few years of a decision to do so. For advanced nations today, possessing the capability for human spaceflight to low Earth orbit is a political and not a technical decision. But going beyond low Earth orbit to the moon is a problem of a different order. And yet the moon is a necessary first step outward for any nation seeking a spacefaring future. So let us look at the resources required to pursue such a future.

The development phase of Apollo required about $80–85 billion in fiscal 2000 currency, about the same as we predict will be required to redevelop similar capabilities. Constellation systems will, as stated earlier, offer substantially more performance than Apollo; but it does seem as if an effort of approximately this magnitude is necessary no matter what. There is an inherent "knee" of the cost versus performance curve; it takes a lot of effort to get to the moon, after which additional capability can be added at somewhat less marginal cost.

So let's assume a minimum required effort of about $80 billion is required to develop a basic lunar capability. In the U.S., at approximate average aerospace labor rates for fiscal 2000 this is equivalent to an effort of roughly 600,000 man-years or 40,000 people for 15 years. Other nations will likely operate in a somewhat "leaner" fashion than is characteristic of the U.S. aerospace culture. I will always remember Max Faget's comment to me that "we could have done Apollo with a lot fewer people, but we couldn't have done it with any more." But it remains likely that an effort similar to Apollo will be required for any nation or society attempting to reach the moon for the first time, provided it has access to the necessary industrial base and an adequate workforce.

Many nations or alliances can as a matter of political choice decide to mount such an effort. Europe has a population 50 percent greater than that of the U.S. yet spends on a per-capita basis only about a fifth of what we spend on space. A future European generation could choose to do otherwise. India has a middle class population equal in size to the entire U.S. population and produces engineering graduates equal to the best anywhere. Chinese space agency representatives have remarked publicly that, today, some 200,000 engineers and technicians are engaged in space-related work. And of course Russia could begin the development of a lunar transportation system today, essentially at its discretion given its existing spaceflight capability and the recent and continuing flow of energy money into that country.

By the mid-to-late 2020s at the latest, several nations will have the independent capability to reach the moon and will be doing so. My hope is that the various programs can be bent more toward a cooperative than a competitive agenda. I believe that nations will find it to be in their interests to cooperate in lunar exploration and development, as they do in Antarctica today. But it will also be true that each nation develop key elements of space infrastructure, especially transportation but also navigation and communications assets, and be unlikely to set them aside in favor of reliance on others. For the next generation, maybe as much as two decades, the U.S. may well be the only nation capable of reaching the moon on its own. But much beyond that, I suspect that we'll be there with others. The moon will be within the grasp of a significant number of advanced nations. It will be the next big leap, a voyage to Mars, where international cooperation is a requirement, rather than an option.

What will be the role of commercial space entities in human exploration? By "commercial space," I mean space business enterprises which develop a marketable capability while dealing at "arms length" with the government, that is, largely without the financial backing and close government supervision which has historically characterized the space industry. The government will, at least initially, still be the major customer for such enterprises. Whether or not an enterprise is part of the commercial space arena depends not on the identity of its customers but on the nature of its interactions with that customer.

I expect that the role of commercial space in human space exploration will be significant and possibly transforming over the next five decades and beyond. We at NASA are presently engaged in an effort to determine whether it is possible for a commercial firm to develop orbital space transportation capabilities without the close supervision of the government. The latter approach through what are commonly known as "prime contracts" with industry has been the traditional approach over the last five decades for state-of-the-art projects in the defense and aerospace industry. It produces successful outcomes with reasonable certainty and at great expense.

I believe it is obvious to most that if a desired product lies within the state of the art, it can be provided with substantially greater efficiency by the commercial sector than by the government. There is little comparative data obtained under controlled conditions to support this claim or to estimate the efficiency factor involved. But to me, the limited data and my own experience points to an efficiency factor of three to seven in favor of the commercial sector. Whatever the factor, the likely cost benefit to the government of commercial procurement of space goods and services, once it is possible, cannot and will not be ignored. But, again, the crucial assumption is that the intended product lies well within the state-of-the-art. When this assumption cannot be met, close government involvement will continue to be required. Commercial firms simply cannot be successful if engaged in a research upon whose success their revenue depends.

Some have opined that the scale and difficulty of spaceflight is such that it will remain an inherently governmental enterprise for the foreseeable future. I do not share this view. For me, the question is more properly "when" not "if" the state-of-the-art in astronautics will permit a private enterprise to develop a successful orbital transportation capability without the direct support—and the accompanying onerous and expensive oversight—of a government prime contract.

We at NASA are attempting to determine whether this date has in fact arrived. By providing "seed money" in the form of Space Act Agreements for two Commercial Orbital Transportation Services (COTS) entities, we hope to stimulate the attainment of entrepreneurial commercial space transportation. If such capability is successfully demonstrated, we can then procure such services in a manner more characteristic of the economy at large than is the usual case in the government-driven aerospace sector. We at NASA are prepared to stand down government systems as and when commercial capability becomes available.

Whether or not the specific COTS initiative is successful, the commercial space business model will eventually become so. A long-term government sponsored space exploration program carries with it the implicit demand for many tons of cargo logistics and crew transport, offering a stable and tempting market niche for industry. Some enterprises will be surely successful in their attempts to service this market and from there commercial space activity will bloom. In addition to transportation, space exploration implies the need for communications, navigation, power systems and other support infrastructure. These requirements will be targeted by specific firms as services to be provided commercially rather than by government.

I believe that the future for U.S. civil space exploration that I have outlined here can be attained with the resources that will be available to NASA by means of conventional government appropriations and acquisition strategies. But I also believe that this is just about as much as we can achieve with those resources unless we can effect real changes in our methods of doing business. If we want to do more, if we want a richer future, if we are unsatisfied by the relatively modest program of inner solar system exploration I have envisioned here, there must be a change in how we go about it. Embracing the possibilities inherent in commercial space transactions is one such method.

What else do we have to do to bring about this future? Most of what we need to accomplish the goals set forth here has already been discussed, implicitly or explicitly, in connection with budgetary issues; but it may be helpful to concentrate some attention on the matter.

The most important factor for future success is stability in purpose, strategy, requirements and funding. Apollo funding was unstable in both directions. The huge rate of early growth allowed the Apollo goal to be met; the abrupt cessation of funding as the goal drew within sight produced strategic damage that remains unto the present day.

To be successful, program managers (whether in government or industry) need stability. Additionally, they need the knowledge that there will be such stability; defensive planning is inherently wasteful.

Stability of purpose, a result of agreement upon priorities, is as important as funding stability. Managers must have reasonable and effective control over what is done with the resources—people, money and time—entrusted to them. If funding is in fact stable, then additional money will not be available to solve problems that are inevitably encountered in any state-of-the-art development program. Managers must have the latitude to sacrifice or defer lower priority efforts in order to protect more important ones. This in turn requires, at a minimum, broad agreement on what those priorities are. When this cannot be obtained, every programmatic overrun and every minor budget variation produces divisive political infighting over what will be sacrificed and what will not. A common result is that nothing is sacrificed and all programmatic content is preserved but at a slower pace. This produces an inherent inefficiency in the execution of all programs, resulting in more overruns, etc., in a degenerating spiral. It is difficult and hugely wasteful to carry out a program in such an environment.

There is another aspect of stability that is equally crucial to bring about in the future outlined here. It involves, once again, a lesson to be gained from the past. This is the absolute necessity of fully utilizing the systems we develop, at huge expense, rather than discarding them in favor of something appealing because it is new. This aspect of stability has had a direct impact on NASA's ability to maintain stability of both purpose and funding for decades.

We must treat our space systems as we have always treated our airplanes. Successful aircraft designs, from general aviation airplanes to the highest-performance military fighters, are evolved, upgraded and used for decades. Just as with DC-3s, B-52s and many other aircraft, we need to understand that Orion and Ares will be flown by the grandkids of the first astronauts who take them into space. We simply cannot again afford the strategic distraction, the wasted money, the squandered talent, and the lost time of building a new human spaceflight system and then using it for only 16 missions.

Once again, a look at the budgetary history provides a sobering lesson for the future, a sobering view of "what might have been." Let's recycle to the early 1970s, a time of budgetary starvation for NASA, a time when we did not yet have the space shuttle but did still have the Apollo systems—the Saturn I-B and Saturn V, the Apollo command/service modules (CSM), the lunar lander and the Skylab system. All of these things were in existence in 1973, having been created in that seminal first 15 years of our agency's history.

Make no mistake; these systems were far from perfect. They were expensive to develop and expensive to operate. Our parents and grandparents, metaphorically speaking, did not really know quite what they were doing when they set out to accept President Kennedy's challenge to go to the moon. They learned as they went along. But what they eventually built worked and worked well. And it could have kept working at a price we could afford.

Let's look at some recurring costs in dollars then and now. All costs include both hardware and mission operations and are at the high end of the range of possibilities because they take no advantage of stable rates of production. Fiscal 2000 costs are approximate, obtained by inflating programs in the aggregate rather than tracking and inflating separate expenditures of real-year dollars.

Element	Real-Year $ M	Fiscal 2000 $ M
Apollo CSM	50	160
Apollo Lunar Module	120	400
Apollo Lunar Mission	720	2400
Saturn I-B	35	120
Saturn V	325	1100
Skylab Cluster	275	925

Let's assume that we had kept flying with the systems we had at the time, that we had continued to execute two manned Apollo lunar missions every year as was done in 1971–72. This would have cost about $4.8 billion annually in fiscal 2000 dollars.

Further, let us assume that we had established a continuing program of space station activities in Earth orbit built on the Apollo CSM, Saturn I-B and Skylab systems. Four crew rotation launches per year plus a new Skylab cluster every 5 years to augment or replace existing modules would have cost about $1.5 billion per year. This entire program of six manned flights per year, two of them to the moon, would have cost about $6.3 billion annually in fiscal 2000 dollars. The average annual NASA budget in the 15 difficult years from 1974–88 was $10.5 billion; with 60 percent of it allocated to human spaceflight, there would have been sufficient funding to continue a stable program of lunar exploration as well as the development of Earth orbital infrastructure.

I suggest that this would have been a better strategic alternative than the choices that were in fact made almost 40 years ago.

After a time, as NASA budgets once again improved, we would have begun to concentrate our lunar activity around an outpost and we would have used cargo missions to emplace the outpost equipment. A modified Apollo lunar module descent stage with extra fuel and cargo replacing the ascent stage could have been used for the purpose. The Saturn V could deliver two such vehicles with a single launch. So over time, we could have built up an early lunar outpost, or smaller ones at different places of interest. By the present day, using what we had with minimal modifications—and I will remind us all that the *Soyuz* systems of that era are still flying—we would have a vast store of experience and a significant amount of lunar infrastructure. When the civil space budget eventually improved, as it did, we would have been well positioned to begin development of a Mars mission. And in the meantime without doubt, we would have continued to modify, refine and incrementally improve the old Apollo designs to the point where they would have provided greatly enhanced effectiveness by the present day.

If we had done all this, we would be on Mars today and not writing about it as a subject for "the next 50 years." We would have decades of experience operating long-duration space systems in Earth orbit and similar decades of experience in exploring and learning to utilize the moon. This essay on "the next 50 years" would be quite different than the one I am offering here. I think most of us will agree that it would have been a better one.

Now, nothing is as easy as planning in hindsight or as permanent as a lost opportunity. I offer the "alternative history" above not to throw stones at policymakers long departed from the scene but to inform future decisions. If we ignore these lessons, we will surely repeat them.

The vision of the next 50 years in space that I have outlined here is not a flight of fancy. It does not require a course change from present understandings or extensive development of costly new technology. It is a logical, incremental, stable and sustainable plan that can be executed with realistically attainable budgets. For these reasons, I believe that it will be done and done as envisioned here. We really can celebrate the 100th anniversary of *Sputnik* with the 20th anniversary of the first human landing on Mars. It is up to us to make it so.